WHAT PEOPLE ARE SAYING ABOUT 'NATURE MAGIC'

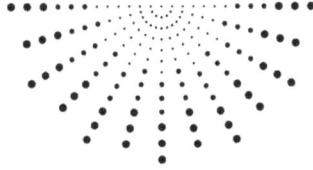

Love how jam-packed and practical this book is. Makes it really easy to see how to get more involved with nature and animals in your everyday life! Really like the ethos behind this book. (5/5)

— TINA, REVIEWED IN THE UK, 28 JULY 2020

Inspiring, insightful and motivating. Well written, a gently flowing easy read. I was transported to a world of nature and abundance in my mind and then inspired to just open my eyes and see that it all exists in my own reality too. Que many private mini nature sanctuaries around the world. And if you like this book you will love the Nature Magic podcast. (5/5)

— MARY BART, REVIEWED IN THE UK, 8 AUGUST 2020

An amazing book for any animal and nature lover with stories and practical tools and suggestions to help get people engaged with the natural world! Love this book!! We need more people doing this work out in the world (5/5)

— INDIE GIRL, REVIEWED IN THE USA, 28 JULY 2020

The author is the director of a well-known nature sanctuary in Western Ireland (Burren Nature Sanctuary) that I recently became aware of, and her new book is great for anyone who is interested in engaging more with flora and fauna. The book is particularly good for educators (that teach children and teens) as toolkits and activities are included. For me, the stories of the animals (including the Emilia the pig who walks on a leash) and her stories of managing the sanctuary (a must-read for anyone considering opening their own nature sanctuary one day) were my favourite parts. I visited The Burren in Ireland in 1997 and it was a beautiful place and had a unique landscape. This book makes my want to return soon! (5/5)

— BRETT, REVIEWED IN THE USA, 28 JULY 2020

Mary's stories are some of the most delightful and enchanting visuals of animals and nature that I have ever experienced. I have my own visual images in my head of Emilia the pig and those images keep me smiling throughout the day. And I've always been scared of goats but now I want to meet Frisky and give her a big hug! And you will want to hug the author, too, for providing this incredibly visual and heartwarming account of falling in love with nature, etc. (5/5)

— KRIS MCPEAK, REVIEWED IN THE USA, 2 AUGUST 2020

This book is a true gem for those who want fun and effective ways to engage people with nature (like hosting a magical pig walk, taking selfies with llamas, and using your goats to teach people about plants!). It should be THE GUIDE for every nature centre and teacher out there. Mary writes from personal experience and holds nothing back about what she and her team have learned since opening the award-winning Burren Nature Sanctuary seven years ago. (5/5)

— REVIEWED IN THE UNITED STATES, 28 JULY 2020

Nature Magic

HOW YOU CAN ENGAGE EVERYONE WITH BIODIVERSITY

Mary Bermingham

FOR JACAN

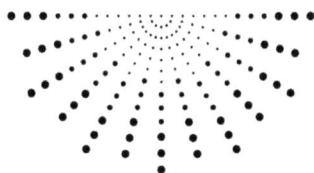

First edition 2020
Copyright © 2020 Mary Bermingham

ISBN: 978-1-8381053-3-4

Editor
Melissa Niemann
Book Layout & Design
Mary Bermingham & Melissa Niemann
Book Cover Designer
Daniela Guedes

The author has written this book based on personal experience and does not claim to be an
expert in the field of animal husbandry, ecology, nature conservation or any related fields. The
advice and strategies found within may not be suitable for every situation and are purely
suggestions.

CONTENTS

INTRODUCTION

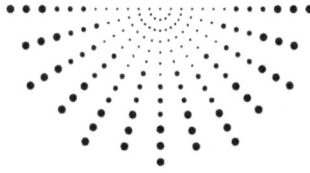

Look deep into nature, and then you will understand everything better.

— ALBERT EINSTEIN

When we turn on the TV or scroll down Facebook and see the Amazon burning or the Great Barrier Reef bleaching, we either suffer paralysing grief and eco-anxiety (described by the American Psychological Association as 'a chronic fear of environmental doom'), or we switch channels and scroll down to escape thinking about it...

But now, more than ever, we must face the facts.

People like Pádraic Fogarty (one of our Nature Magic podcast guests), author of Whittled Away: Ireland's Vanishing Nature; and Catherine Farrell, an ecologist at INCASE (Irish Natural Capital Accounting for Sustainable Environments) are at the forefront of helping people see and understand the facts about our natural environment.

But how do we address these facts? We need to hold a vision of what we want to see and ask ourselves: *If I had a magic wand, what would I like the Earth to look like right now?*

I imagine a paradise on Earth where there is room for all the creatures and plants, living in harmony with humanity. I imagine the forests and oceans alive with biodiversity... A world where we have put the carbon back in the ground, stopped producing plastic and collected every last bit of it; where governments have banned herbicides and pesticides, and organic farms produce abundant, wholesome local food for their communities.

Is this a pipe dream? I don't think so…

Over the last few months, during the Covid-19 crisis, we have seen how everything can turn on a sixpence, how humanity can make changes overnight. We need to end the dominion over the plant and animal kingdoms and usher in an age of respect and cooperation.

In the words of Costas Christ, "There is no separation. I hope as we come out of the Covid-19 pandemic, we will understand that we cannot have personal health and wellbeing when we don't have planetary health and wellbeing."

So, what about the people who don't understand the crisis, or those who don't care? What about people in cities? Or people who believe they need to use up Earth's resources to survive? We need to show them (as Jane Stout from the Irish Forum on Natural Capita does so clearly in her interview) the value of nature—not only to ourselves but as the cornerstone of our economies. And we need to get them out of the cities and introduce them to the natural world.

I believe the first step to combating the climate and biodiversity crises is to engage people with nature and create a nation of conservationists. For example, in the words of Patrick McCormack, "I often say as a farmer, the biggest drawback I have, is that I'm dealing with a public that's 95% ignorant of how food is produced."

If people were better informed, they would not accept the Teagasc's (Irish semi state agri-food advisory service) advice to coat crops with highly toxic Glyphosate. If they could meet Emilia, our pig, they would not want to eat that slice of ham from an animal that had lived locked-up in a small cage for its entire life. If they had to keep a pig in their own garage for six months, confined to a small crate, could they live with themselves? Of course not! That pig would soon be up on the sofa and hopping in the car to go for walks on the beach; because people are kind.

The transparency of the internet and social media has awakened people to reality.

What I hope to achieve with this book is to offer easy, fun solutions to the ever-pressing problem: how to engage people with nature. We have learnt many ways to achieve this through running The Burren Nature Sanctuary for the last seven years.

At Burren Nature Sanctuary, our vision and mission are:

VISION

- To allow people of all ages to engage with the unique biodiversity of the Burren.
- To foster a love for nature.
- To inspire people to cherish, respect and conserve our natural heritage for a sustainable future.

MISSION

- To introduce people to the Burren by providing an accessible place to engage with nature.
- To build on the Living Collection of Burren Flora, Trails and Exhibits.
- To educate people in nature conservation through an enjoyable experience.

If you bring people into nature, they will love it. They will become interested in conservation because every one of us cares deeply about the things we love. But you cannot fall in love with someone you haven't met… By encountering flowers, trees, animals or a natural landscape such as the Burren, people get the time and space to fall in love with nature.

As Patrick McCormack says, "You can't buy the 'experience', you have to let nature *flow* into you."

The Burren is a unique and magical place. Mountain avens from the Arctic bloom next to orchids from the Mediterranean and maidenhair ferns from the tropics. Lakes disappear and reappear daily like clockwork. Fossils of ancient tropical sea creatures decorate the limestone and mountain goats leap from boulder to boulder. Otters play around the ruins of castles perched at the edge of the Atlantic. Rainbows shoot across the hills. The pink sun shines underneath dark, grey banks of clouds that slowly sail around, watering this garden of Eden. It is no wonder Tolkien used the Burren as his inspiration for his books, and that this landscape draws mystics and poets from all over the Earth.

We want people to witness this beauty and fall in love. We hope to teach by osmosis. We hope the visitor leaves with fond memories of a fantastic day out, knowledge about the diversity of the region and how to visit this fragile landscape mindfully, and a lifelong bond with nature.

This book outlines the strategies we have devised over the last seven years to entice and engage people with nature in fun, exciting and innovative ways. People come for a fun day out and leave armed with knowledge. Anyone can replicate the advice, activities and strategies we share: from policymakers, to biodiversity officers, to teachers, to parents, to eco-warriors.

At the end of each section is a 'Magical Story'—a personal anecdote about connecting with nature that will hopefully entertain and inspire anyone that just happens upon this book.

There is also a brief section on tips for opening a nature sanctuary. (Please contact me for further advice if you are embarking on this road or if you have questions at https://www.naturemagic.ie/contact)

The Nature Magic Podcast

During the Covid-19 lockdown, we started the podcast Nature Magic: A Positive Voice for Nature, where we interviewed influential people, all working hard to create environmental awareness. The questions were simple:

- How did you become a nature lover?
- What is your favourite plant or animal?
- Do you feel spiritually connected to nature?
- What can people do to help nature?
- If you had a magic wand, what is the one thing you would you do to help the planet?
- What is your favourite nature book?

The decision to speak publicly was hard. I had no experience as an interviewer. It was scary but incredibly rewarding. The conversations needed to be had, and the voices needed to be heard. To my delight, the podcast charted at the number one position in the nature category in Ireland in a very short time! (Check it out at https://www.naturemagic.ie.) Before each section, you will find inspirational excerpts from these valuable and passionate interviews.

How I became a Nature Lover

When I was 14, I read The One Straw Revolution by Masanobu Fukuoka and became an instant nature activist. I volunteered for Friends of the Earth at age 15 and worked in their London office (which led to some interesting family conflict!). Although my life veered around all over the place (I trained three-day event horses and then studied and worked as an engineer), I never lost my love and respect for Mother Nature.

My husband (Roy) and I opened the Burren Nature Sanctuary in the post Celtic Tiger recession in 2013. I had experience with design and planning, and Roy had experience with construction and project management. Our skills complement each other, but that's not to say it's always easy working with your partner!

During planning and construction, only three local people had faith in the project (apart from Roy and I). They were Kieran Whelan, who laid the blocks for the café; and Bob and David Flaherty, who (somehow) created our lovely nature walk by weaving through a mile of shattered limestone rocks, with nothing but a tiny digger and motorised wheelbarrow. We appreciate your support!

Everyone else said we were crazy. Then, when we opened, more critical comments followed:

Why do you need a sanctuary for bushes?

Nobody will come to look at a flower…

But in these few short years, people's perceptions have transformed. For example, the concept of devoting half the farm to re-wilding is now more widely understood than a few years ago. Our local community now loves Burren Nature Sanctuary, and we enjoy support from all corners of the Earth.

Find all our free bonus material at www.naturemagic.ie/bonusmaterial

Contact Mary at www.naturemagic.ie/contact

Nature Magic

PART ONE
ENGAGE WITH NATURE

NATURE LOVER INTERVIEW: PATRICK McCORMACK

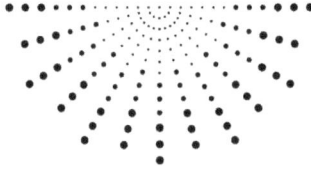

If you want to sow seeds, you have to prepare the ground.

— PATRICK McCORMACK IS A POET AND PASSIONATE ORGANIC FARMER IN THE BURREN

HOW DID YOU BECOME A NATURE LOVER?

Farming is my passion. It's what I do. It's all I know. I cherish being alone in the Burren; I cherish the times I have spent alone in those hills.

WHAT IS YOUR FAVOURITE PLANT OR ANIMAL?

I think one of the wonderful things about being in the Burren, when we herd cattle in these mountains, is the blue grass.

The old people used to say, when the blue grass would come, the worst of the winter would be over. And, of course, then we're all waiting for the gentian (a flower). The gentian, my God, you'll see that incredible blue coming out of the Burren, because the Burren is shorn of its colour at that time of year. There is nothing—only brown and dark—and then you see this most incredible blue. The mystics tell us that is the first colour we see when we pass over to the other side.

The next thing that will arrive is the early purple orchid, this is the most incredible thing, each wave of flowering. The burnet rose is my favourite; it blooms for two weeks. I also like the Potentilla down by the turlough, the hawthorn blossoms and the blackthorn. Just to be part of the spring rising, the colour, the energy, the celebration of life, in itself, is marvellous.

I always say the peregrine falcons are my life's greatest teachers. They soar so high in the sky; their vision is 15 times better than ours. They migrate to the coast for the winter and then, in the springtime, you hear this cry—this eerie cry. The falcon has returned! I feel a rush of blood just thinking about it.

In the biggest challenges of my life, they would always appear. And I know the message they came with was, "It's going to be all right."

DO YOU FEEL SPIRITUALLY CONNECTED TO NATURE?

A day doesn't go by that, if I was by myself trying to herd cattle, I'd make it up to the old fort where the graves are and ask, "Come on lads, any chance of a helping hand?"

You see, there are other ways of doing things!

WHAT CAN PEOPLE DO TO HELP NATURE?

How can we shift the focus? Because the whole education system is all geared to take something from the planet, to be successful in monetary terms or for careers. So, how can we shift the focus to putting something back into the Earth?

I don't have the answers…

In the words of Bryan McMahon, "No idealistic goal can be achieved through fear. It needs to be inspired by some kind of contagious, dynamic love."

IF YOU HAD A MAGIC WAND, WHAT IS THE ONE THING YOU WOULD YOU DO TO HELP THE PLANET?

> Plastic is our gift to the world, isn't it? Our generation's gift!
>
> Get rid of plastic and stop spreading nitrogen and raw slurry on our fields. That is just bad science.

Visit www.naturemagic.ie (or search 'Nature Magic' on all major podcast platforms to listen to the full interview).

DOMESTIC ANIMALS

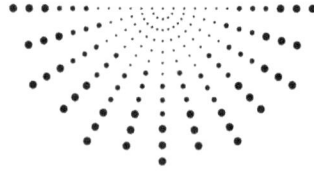

INTRODUCTION

Animals are the easiest way to connect with nature. Who doesn't want to save the elephant or the panda? Not so many people want to save the wild chive… Who doesn't want to pet a rabbit, take a selfie with a llama or walk a pig?

In this section, you'll learn which animals are happy to take part in interactions with you. It explains how to run fun activities, how to manage and care for animals and the lessons animals can teach us. As I introduce you to the animals, you'll see a rating for each based on their engagement level, ease of care and expense to keep.

I hope this section will be of help to those who are considering adopting, for example, a pet pig or an alpaca. Furthermore, this section is a resource for nature centres.

Animals attract people and teach them to connect with nature. If, for example, someone comes to take a photo of the alpaca, they might join a guided botany walk or herb workshop. You will soon see them falling in love with the trees and plants they used to pass by as if they didn't exist.

Not everyone can keep a pig or an alpaca, but nearly everyone has room for a pair of guinea pigs. These adorable, fluffy pets are the most accessible entry point to pet ownership, and they have a five-star engagement rating!

Always wash your hands after animal interactions. Make sure you have a dedicated hand washing station near the animals.

THE FAIRY PIG WALK

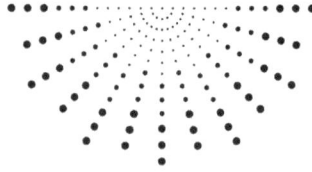

```
Animal Engagement Rating:

★★★★★

Difficulty of Care Rating:

★★★

Expense of Care Rating:

★
```

INTRODUCTION

Pigs are enchanting. No one can resist them. They are a shoo-in to dupe the most disengaged individuals into spending time in nature.

People have dragged the dourest of city slicker husbands and the grumpiest of techie teenagers to the Fairy Pig Walk, and Emilia has never failed to 'turn that frown upside down!' People come from far and wide to do the Fairy Pig Walk. And Emilia gets stellar reviews!

" This is the best thing we did in Ireland.

" This is the best thing we did on our trip.

And even,

> This was the best thing I did in my life!

Ninety-five per cent of our Fairy Pig Walkers (booked through Airbnb experiences) are adults.

We also run a children's pig walk on Sunday mornings—generally with a large entourage of little 'helpers'.

A lot of our visitors who go on the Fairy pig Walk are considering owning a micro pig. As with every animal chapter, I have added sections with advice about care and maintenance to help you make an informed decision before you buy a pig.

Pigs are a joy to keep, but as with any animal, they require a serious commitment. We often get calls to take in people's micro pigs because they have become bored with them, or the pigs have grown too big for them. Sadly, many pet pigs suffer unhappy endings, as sanctuaries adopt only a lucky few. So, if you are not entirely sure whether ownership is the right route for you, just support your local sanctuary, visit a pig like Emilia, or read on!

My friend, Helen, ended up hosting our most heart-warming Fairy Pig Walk. On this day, only two people had booked to go on the walk. They were a young couple from Seattle, Washington. The woman was desperate to own a pig; her partner… Not so much! Of course, her tactic worked, and he instantly fell in love with Emilia. As they walked through the woodland, the conversation went like this:

"Well, your place is too small…"

"And your place doesn't have a garden…"

They laughed and took photos as they skipped from Fairy Door to Fairy Door with Emilia. Helen shared some history and told them folklore from the area. She also handed them the beautiful poem, The Song of Wandering Aengus, by W.B. Yeats (see page 151).

By the time they got to the Fairy Wishing Well, they were utterly enchanted and had decided they would move in together and get a pig!

He dropped to one knee…

"Will you marry me?"

"Yes!"

High five, Emilia!

Helen has also been persuaded to be Biddy the Witch and Mother Owl at Halloween, Mrs Claus at Christmas and even the giant chicken for 'Chicken Run' at Easter. She quit the giant chicken job after a few rugby-tackles by overzealous egg collectors. These days, our giant chicken always has a bodyguard!

Her most stressful walk was with a group of ten Americans. When hosting the Fairy Pig Walk, you have an enormous responsibility to give everyone a wonderful time.

The group had flown to Ireland, got to Galway, booked their tickets, got to Kinvara, found the Burren Nature Sanctuary on time, signed in, got their coats on and were prepared to enjoy themselves.

They got to Emilia's paddock with much excitement, and… No Emilia. The gate was swinging wide open. Someone had let her out. Helen divided the group and sent them on a search-and-rescue mission. When they finally found her in the woodland, there was jubilation. Helen had to recuperate by having a big lie down in the office afterwards!

Later in this chapter, I will share a few of the hundreds of five-star reviews Emilia has earned. If you want to see some lovely photos of her, or book the Fairy Pig Walk, find her at https://www.airbnb.ie/experiences/446234, or just Google 'Fairy Pig Walk' (there is only one!).

EXAMPLE ACTIVITY: THE FAIRY PIG WALK

Time: 1 hour

When Emilia was one, she started taking children on tours of the Fairy Woodland. She was already used to the harness, enjoyed the soft surface of the woodland and loved people of all sizes. When Airbnb started their experiences, we uploaded cute photos of Emilia and the bookings began rolling in.

Number of People: 1-6

We prefer to work with groups of up to six people. When we first started, we tried to accommodate up to ten. It was nearly impossible to give each member in the group time with Emilia. These days, we will only take on larger groups if it is one family.

Experience Outline from Airbnb

Arrive at Burren Nature Sanctuary and receive your admission bracelet and a wooden token for tea/coffee/hot chocolate (to redeem when you wish).

Also receive, as a souvenir, poems by W.B. Yeats who researched his famous book, Fairy and Folktales of Ireland in the local area.

Meet and feed the farm animals: Frisky the goat and her friends Hannah, Mocha and Chino; Jazz the alpaca, Frank the kissing llama and Ella and Lola the rabbits.

Continue on to meet Emilia the Juliana pig, who lives with Dyson, the not-so-micro rescue pig!

Emilia will run out of her paddock. Mary or Helen will put her harness on; and you will learn how to lead Emilia and how to use the pig paddle to direct her.

Set off for the nature walk.

Learn about the famine village. Emilia is not a fan of history, so she won't hang around there too long…

Lead Emilia around the Fairy Woodland and learn who lives in the Fairy Houses. Next, make a wish at the Fairy Wishing Well.

Take photos with Emilia next to the see-saw and share them on social media (or save one as a Christmas card image).

Return to the café and gift shop or continue on the mile-long nature walk, using the free downloadable Burren Nature Sanctuary audio app as your guide.

Browse local and artisan products in our AVOCA shop and, if you wish, have lunch in the Burren Nature Sanctuary café with locally sourced fresh, delicious ingredients.

Check out our online gift shop: https://www.burrennaturesanctuary.ie/shop

Airbnb is a wonderful platform for exposure. You get instant worldwide reach, especially if your experience has a high rating. Experiences, like The Fairy Pig Walk, pop up whenever people search for accommodation in the wider area.

EXAMPLE ACTIVITY: SUNDAY MORNING PIG WALK WITH CHILDREN

Time: 30 minutes – 1 hour

Number of people: no limit

We have had groups of over 30 people on Emilia's Sunday walks.

Experience Outline

Emilia runs out to the bench to put on her harness, and I introduce her. Typically, children and parents who meet her for the first time will stand back at a respectful (and sometimes fearful) distance. I encourage the children to interact with her:

"You can come over and rub Emilia anywhere you want."

"Isn't she soft?" (She isn't!)

"No, bristly!" They shout.

"In the olden days, they used to use the pig's hair for paintbrushes and even toothbrushes!"

That always gets a good response!

"Emilia's favourite thing is food and her second favourite is tummy rubs!"

Then I point out her features:

"Emilia is a Juliana pig. They have a straight tail and a short snout, so they don't dig up your garden."

She is wagging her tail at this point, happy with all the attention.

"Pigs wag their tails just like dogs when they are happy. We are going to take Emilia around the Fairy Woodland, and there is only one rule… Emilia is the leader! Who is the leader?"

"Emilia!"

Small people have little patience, so the introduction is short and sweet. I pass the leash to the nearest child, and we head off slowly. As we walk, I stay in front of Emilia with a scoop of feed and the pig paddle. (When we do these walks, she only gets a handful of breakfast. This prevents her from being over-fed.) If I see someone lagging, I will put a little feed on the floor to keep her busy for a while. If anyone tries to pass her by, I will re-mind them of the rules. Any size group is manageable if everyone stays be-hind Emilia during the walk; otherwise, total chaos can descend.

The Sunday morning walks usually comprise 20 to 30 kids and parents. I try to let everyone, even the smallest toddler, have a go at holding Emilia's 65-foot (20 m), retractable leash. Emilia doesn't care if you accidental-ly drop the leash and it drags after her. She's not going anywhere without you. Emilia knows her route! Even if you lost her, she would eventually wander back to her paddock for dinner.

As one German man said in a review, "The pig doesn't need walking, she knows the way herself."

I stop in front of each Fairy Door, and Emilia enjoys her breakfast while I tell the children who lives in each Fairy House, what jobs they do and how they make Fairy Dust…

Check out the Fairy Rules and Jobs in our bonus material: http://www.naturemagic.ie/bonusmaterial

SOME OF EMILIA'S FIVE-STAR REVIEWS

> Visiting the Burren Nature Reserve was a special experience. It felt very natural and the animals seemed to be very well kept. Meeting and walking with Emilia, helped by Helen, was definitely the highlight. Helen was really helpful and friendly and Emilia is a real personality—great fun!
>
> — JOANNA, AUGUST 2020

> That was one of the best things we have ever done :-) An amazing experience. You have to do this, too. Thanks for everything!
>
> — SEBASTIAN, JUNE 2019

> Such a weird and wonderful thing to do in Ireland, I went with my girlfriend and the experience was amazing. The animals are so well looked after and you don't get the opportunity to walk a pig anywhere else!
>
> — MATHEW, FEBRUARY 2020

> Such a wonderful experience! No matter what I'm going to say, it won't be enough to describe it. You better come and check it out, it would be something you won't forget. Mary,

(the host) you are amazing and you make this place and experience even more magical!

— INA, FEBRUARY 2020

This experience was so, so worth it. It was our favourite thing that we did while visiting Galway. We laughed from start to finish and just had the most perfect time. I would definitely recommend this to anyone visiting Galway.

— DANI, FEBRUARY 2020

Walking with Emilia and touring the sanctuary was such good, pure fun! Emilia's giddiness over her walk is contagious! The simple joy of getting to feed the goats and pet Frank (the llama) and the other animals was such a refreshing break from busier city life!

— LOUISE, OCTOBER 2019

The entire sanctuary is a very special place. We have enjoyed every minute of it. The pig walk with Emilia was really fun! She is full of energy and showed us all the fairies. We would definitely go again on this experience.

— KAT, OCTOBER 2019

Such an incredible experience, one of our favourites in Ireland! We (two adults) had such a fun time walking with Emilia through the forest and being able to interact with the other animals on the property. Mary had great insight into the natural surrounding and history on all the animals on the reserve. A great activity for any animal/nature lover!

— EMILY, OCTOBER 2019

> You get to walk in a fairy forest... With a pig! What's not to love?

> — LEIGH, OCTOBER 2019

> This experience was the best part of our trip to Ireland.

> — VANESSA, OCTOBER 2019

> This was the best experience we have had on our trip!

> — LIZ, SEPTEMBER 2019

> What more do you want? A pig in a magical fairy forest.

> — ART, AUGUST 2019

> This was such a special and outstanding experience.

> — DIANA, SEPTEMBER 2019

> Such a special experience! Absolutely magical!!

> — ROSE, AUGUST 2019

> So. We walked a pig through a fairy forest. It was fantastic. The End.

> — HEPHZIBAH, AUGUST 2019

> Amazing walk with a pig!

> — MANUEL, APRIL 2019

> A "pig walk" was an unique experience, thoroughly enjoyed myself.

— JOE, MAY 2019

Joe was our oldest guest on *The Fairy Pig Walk* to date. He is a US war veteran with a prosthetic leg and was 89 years old when he did the walk. He goes on adventures in memory of his late wife. There are quite a few roots and stones on the trail. I was glad Helen was there to link arms with Joe in certain areas. We all agree: Joe is a total legend.

> This is a beautiful experience for everyone's inner child.... Helen was a wonderful hostess and we so enjoyed this moment to walk in the Faerie Wood and step back to a simpler time. By the way, we are in our 70s.... Never stop walking with the fairies.
>
> — ALICIA, JUNE 2019

> A heart-warming experience to boost anyone's mood! So brilliant.
>
> — LEAH, MARCH 2019

> Great experience. Never thought you could walk a pig!
>
> — MARK, APRIL 2019

> Great fun for all ages. All the animals are very cool.
>
> — CAILEAN, AUGUST 2019

> This experience was beyond amazing.
>
> — CATHERINE, SEPTEMBER 2019

There are a few four-star reviews (booooo!), including: *"The pig is too big."* (Sizest!) And, *"The pig is too fast."* (To be fair, she was very fit at the time!)

Lessons Our Visitors have Learned from the Fairy Pig Walk

- People develop a love for animals when they meet Emilia.
- They learn that walking in nature is fun. They learn about native trees and plants and how people from the famine village used to live off the land and sea by eating berries and oysters, and how they used the hazel trees for fencing and roofing.
- They learn that the woodland gives them a sense of calmness.
- People develop a connection with Emilia, the woodland and the rest of the group.

Lessons We have Learned from Our Pigs and the Fairy Pig Walk

- Smaller, intimate groups of adults work best.
- Emilia doesn't enjoy the rain, so we don't run The Fairy Pig Walk from November to February. She will only put up with light showers and refuses to come out of her ark if it is too wet. We state in our terms that Emilia has the right to cancel the walk if there is bad weather!
- We have learned to ask people to come at least half an hour early as they usually come late.
- Emilia likes the wood-chip path in the woodland. Walking on tarmac or a path can make a pig's feet sore. If you think it is easy putting four shoes on a pig, think again… We tried it with our first pig, Norman. It didn't go well.
- Give people more than they expect. Our guests get to enjoy the rest of the sanctuary and the other animals for as long as they want. They get a free hot drink and receive a print-out of a beautiful WB Yeats poem to take home.
- If one kid is going along with a group of adults, it can get tricky. They may want to take over and guilt-trip the other adults into letting them hold the leash too much, and their parents encourage them to be the star of the show! Assertiveness is necessary. Each time Emilia stops, you may need

to pass the leash around without asking. If a toddler is leading the pig, an excellent phrase to use is, "You are so good at taking turns." Then hold out your hand for the leash. This nearly (nearly) always works.

- People on the nature walk, not booked in for The Fairy Pig Walk, try to join the group. This can be embarrassing, especially in the busy season (when more people are walking in the woodland). I usually stop so people can say hello to Emilia and then explain that we are in a private group session.
- Most people who book The Fairy Pig Walk are animal lovers already. This makes it easier to get them engaged with other aspects of nature.
- Paying before the visit works well as the memory of parting with money has faded by the time guests arrive for the experience. It is also much easier for our staff to give visitors the 'Disney' customer welcome if they are simply greeting guests, not taking payments.
- People are generally giddy with excitement when they arrive, which makes for a fantastic start to the experience.

PIG CARE AND FACILITIES

Shelter

Pigs need a good shelter. They can catch a cold if exposed to the elements, and it can become life-threatening. Our pigs have a timber pig ark full of good quality dry straw. The entrance faces away from prevailing winds. Make sure your shelter is tightly secured to the ground so it can't blow away. If you cannot provide adequate outdoor shelter in bad weather, your pig(s) must be brought in to a stable and exercised daily.

Pigs need as much space as possible. Our paddocks are about 65 ft x 65 ft (20 m x 10 m). They will root under the most secure fence and escape. So, if you want to keep them in their paddocks, you will need an electric fence about 6 inches (15 cm) off the ground. They will also need a secure gate. Pigs need clean water, and because they like to root, they will need their water trough/drinker cleaned out regularly.

Food

Pigs are omnivores (they can eat vegetable matter and meat). They like to forage and graze.

Although they can eat meat, we don't feed it to ours. We feed them grain and any vegetable scraps we have. They need a varied diet and will eat grass and root around for worms and snails, etc.

With most pigs, especially small breeds, the main concern is not to over-feed them. We keep a careful eye on body condition and restrict rations if they get too heavy. If pigs are overweight, it is uncomfortable for them to walk, and they will avoid it. This may become a vicious cycle. Ask an experienced pig owner or vet if you are unsure about your pig's body condition.

The key to having pigs that are safe with people is to NEVER (from the time they are born) feed them snacks from your hand(s). Feed your pigs on the ground or in their feeder at regular feeding times. Our two pigs are an excellent example: Emilia has been with us since she was a piglet and is happy to say hello to anyone. She loves being scratched and rubbed; she never bites. Dyson, who had been spoiled rotten before he came to us, is always nibbling at your pockets, looking for snacks.

Enrichment

Pigs love to wallow in mud; it helps them to cool off in hot weather. Dyson loves a muddy puddle, Emilia prefers to paddle with Helen in her paddling pool.

Lastly, pigs love to scratch. Fix a yard brush head on the outside of their sty, and it will delight them.

Health Issues

If there is something unusual about your pig's behaviour, for example, he/she refuses food, pants or has runny eyes, call the vet immediately. If your pig looks sick, take this seriously. Pigs can get chills and colds just like humans, especially when the local weather is changeable and they live outside. A pig's health can deteriorate quickly.

Sun exposure: our pigs are not pink and do not seem to suffer in the sun. But pink pigs get sunburn and need careful monitoring. You can apply sunscreen.

Pigs need to be de-wormed at least once a year. Ask your vet about this.

Adopting a Piglet

Emilia is a Juliana pig, which is a small breed. I highly recommend them. She is about knee-high, has a short snout and doesn't dig, which is a massive bonus if you don't want to turn your grass into a ploughed field!

Emilia loves people.

All piglets are small, but there is no guarantee that they will stay that way. This is true even for small breeds. There are lots of 'mini' or 'micro' pigs for sale. They will not stay 'mini'; and as for teacup pigs, they may exist, but I have never seen one…

Pigs do all their growing up during their first year, so make sure you see the parents. When you bring your piglet home, take her outside and wait for her to do her droppings. After that, she will continue using the same place and will never mess in the house. You only have one chance at this.

If pigs don't want to do something, they will let you know in a high-pitched scream! Piglets naturally hate having their hooves off the ground and will squeal like crazy if you pick them up. They will get used to it if you insist, but it is natural for them to be afraid of this.

They love lying by the fire and snoozing around, but they can also make mischief.

Emilia's sister lives in Ennis, a large Irish town. Her owner rang to ask us why she was taking the clothes out of the washing machine and pulling the curtains down. She was probably bored or interested in making a comfy bed… So, you can see it might be a little awkward to live with someone like that.

Adopting a Rescue Pig

We adopted Dyson, our other pig, when he was already fully grown. A lady in Kerry (Ireland) bought him as a 'micro' piglet.

There is nothing 'micro' about him now.

Her son begged us to adopt him as he had taken over the house, dug up the garden (he has a very long snout like a wild boar) and chewed the car seats. He had even started 'barking' at visitors and had become quite protective of his owner, chasing away anyone who wanted to visit… Even the postman was in trouble!

As he is a boar (a boy pig), he had also grown two large protruding tusks (teeth) and looked quite menacing. (He would use them as weapons in the wild, and they continue to grow throughout their lives.)

When we collected Dyson, we were told: "He only eats tins of fruit salad with syrup, organic chocolate and cheese strings; and he drinks Capri-Sun." (A sugary, orange kids' drink with a straw…!)

Dyson was living the good life, going to the beach in the car and being the boss of the household. This lady loved him TOO much! The consequence was that she couldn't manage him and had to get rid of him.

We tempted Dyson into the trailer with pieces of organic chocolate, drove him across the country and let him into the stable. He was desperately sad, lonely and homesick. He curled up facing the wall. He stayed there for 48 hours and refused to eat. We left oats, diced fruit and vegetables for him. Eventually, he accepted the high life was over and he ate his vegetables!

Because of his bad habits formed in his early life, Dyson cannot interact with the public. He is continually looking for treats and will run to you, chewing at your trousers (which can be quite alarming with his two long tusks).

He will go for a walk with me off the leash, but we can never really trust him with other people. During the Covid-19 lockdown, I allowed Dyson out to have a ramble for himself every day. One day, our neighbours came over for a walk with a group of French students. They met Dyson in the Fairy Woodland and got the fright of their lives!

Dyson's life has changed, but he now has the freedom to roam around and use his big, long snout to dig as much as he likes. He has destroyed his paddock, there isn't a blade of grass growing, and he has dug up stones the size of basketballs. He now has a healthy diet and a companion of the same species, so he is one of the lucky ones.

We get many calls to take in pet pigs that people have become fed up with. But each pig needs proper facilities, so sadly, we cannot take them all.

Training Your Pig

Get your pig used to a harness as soon as possible. The best way to put it on is to put some food on the ground in the same place every day and clip on the harness while she is busy.

It is more convenient to use a retractable dog leash, but you can also train with a short leash (she will get used to it, eventually).

Do not try to catch and wrestle with your pig… She will win!

Equipment

A **Pig Paddle** is handy for training and can be made at home with little effort. Take a piece of board of about 12 inches x 20 inches (30 cm x 50 cm) and attach it to a stick that reaches up to your hips. The pig paddle can then be used to move your pig along from behind (as you would see people walking around a show ring, directing their pigs with a large board). You will only need to touch her (or push a little at times).

Also, use it to change direction by placing the paddle to the side of her face, in the direction you do not want her to go.

The pig paddle is your brake, accelerator and steering wheel.

When your pig is trained a little, pull on the leash to slow her down. If you take her morning feed with you and put a little on the ground every time she goes in the right direction, she will quickly get the hang of it.

Try not to pull on the lead too much as a pig's skin can quickly get rubbed and sore.

It can be a juggle at first with leash, food and pig paddle, so it is handy to have an assistant. But after a while, your pig will happily walk alongside you.

Pig Clothes: we don't put clothes on Emilia, but she wouldn't mind. Follow Esther the Wonder Pig on Instagram to see all the outfits she has! @estherwonderpig

BONUS FUN WITH PIGS

It is Easy to Pull a Raffle with a Pig.

Lay all the tickets out on the floor and put a strawberry on each one. The first strawberry the pig eats is the winner. We did this at the Discover Ireland Tourism Centre during the Arts Festival in Galway City. At the time we still had Norman, our first pig. He was a fantastic character (RIP). It was such a success that a huge crowd followed him like in the Pied Piper back to the car.

I knew Norman would scream when I tried to pick him up (not an easy job as he was fully grown) to put him back in the car. So, I waited for everyone to go away. They didn't.

They were utterly shocked when they had to witness me wrestling with a pig who was struggling and screaming like the world's most furious toddler. As soon as he was in the car, he was ok, and his audience breathed a sigh of relief!

Cone Slalom

One of our trickier age groups is teenagers. Only a handful would call themselves nature lovers; they mostly want to stare at their smartphones.

Galway County Council had selected and funded us to run a biodiversity course of three days throughout the year, in different seasons, for Transition Year students (15 to 16 years old). We called it Cycles of Nature (see page 194).

The teenagers were walking up from the village, too cool to wear any coats; most of them were wearing white runners. There were 60 kids in the group. Only a handful were interested in finding out which flowers were blooming in which season.

We did an animal interaction every day as an icebreaker, and it was a resounding success. On the first day, we had put cones along the field and divided the group into teams. We handed them a bucket of food and the pig paddle. Then we told them to get Norman to slalom (zigzag) through the cones. Teenagers fell over, ruined their sparkly white runners, collapsed into hysterics; and lots of photos were taken.

It is not easy to steer a pig! Conversation flowed, the group bonded. Everyone was happy.

ANIMAL HANDLING: GUINEA PIGS

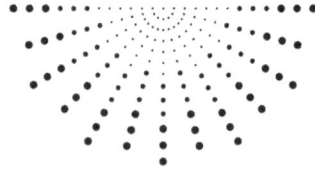

Animal Engagement Rating:

★★★★★

Difficulty of Care Rating:

★

Expense of Care Rating:

★★

INTRODUCTION

Guinea pigs are our top choice for an entry-level animal interaction to engage people with nature. These animals are sweet-natured and easy to care for. They quickly learn to sit happily in one spot, and this allows people to stroke them. They do not bite (unless you jam your finger into one's mouth, in which case he might think it's worth tasting), and they rarely scratch.

It is easy to show someone how to pet a guinea pig gently, and it is especially fun to see when people realise a guinea pig can 'buzz' (like a purring cat) when it is happy. Dave, our first guinea pig, was always so excited to get his two daily 'massages' from visitors. At animal handling time, he would call out when he saw people coming into the barn!

EXAMPLE ACTIVITY: HOLD A GUINEA PIG

Time: 1 - 5 minutes per person

IMPORTANT: Never feed guinea pigs while doing this activity. In fact, it's best not to feed guinea pigs at all when they're being handled. That way, they will learn not to expect food whenever you are holding them.

We do this activity twice a day at Burren Nature Sanctuary (at 11:30 and 15:30) in high season. We have had up to 30 children and parents with one (experienced) leader at 'animal handling' sessions.

To run this activity successfully, you will need to practise daily with your guinea pig. The key to success is for your guinea pig to experience the event the exact same way every time.

If someone else also runs the activity, make sure they receive training to follow the outline with precision.

Activity Outline

Cut an old carpet into squares of about 16 inches (40 cm).

Have a pair of guinea pigs in a cage with a top-opening lid. Get everyone to sit in a row on a long bench. Place the carpet squares on their laps. The carpet squares protect clothes from occasional poop and little children's delicate skins from the guinea pigs' nails.

If children are taking part in the activity, ask the following question before you start (always use positive language when offering instructions):

"What kind of vegetable does your finger look like?"

"A carrot!"

"Correct! What would happen if you put a carrot in a guinea pig's mouth?"

"They would eat it."

"Correct!"

Guinea pigs will only bite by mistake if they are expecting food, or if you get in between two males having a squabble. Over seven years and thousands of children, (touch wood), no child has ever been bitten. (If this does happen, write an accident report, advise a trip to the doctor and offer to pay the doctor's fees.)

Place the guinea pig on the carpet, pointing in the direction that they are going. Young guinea pigs may move around, so pick a fully grown guinea pig and practise with him until he is used to sitting on the carpet. Always go in the same direction along the line of children to avoid confusing the guinea pig.

Place the guest's two hands on top of the guinea pig with your hands. Place one hand on the rump of the guinea pig and explain how that hand stays still. The front hand can then stroke the guinea pig in the direction of his fur, from head to back. Children may try to pick up the guinea pig —simply explain to them that the guinea pig has been trained to stay on the carpet, and this is what he likes.

Move along the line with the guinea pig. When you are confident with your guinea pigs, you can have two going along at different places along the route. That way, each person's turn with a guinea pig is longer. (Or, ideally, have two members of staff leading in the activity—one for each guinea pig.) After a few minutes, (a few seconds for tiny children) move the guinea pig along to the next carpet square and place it in the same direction, looking towards the route. For extra fun, people can get a photo of their children holding the guinea pig.

Boys or girls?

We find girl guinea pigs to be quieter than boys, but our first honorary 'member of staff' was a male black and white guinea pig called Dave, and he was a total cool dude.

He was so popular that the same kids used to come back again and again to visit him. I would tell them stories about what Dave did in his spare time: that he had a little leather jacket and a motorbike and liked to go to the village with his friends; or take road trips around Connemara. The parents would crack up at all the antics Dave had been up to.

When Dave passed on, it was a hard blow for the whole team.

Our next hero was Pippin, a very fancy long-haired Peruvian, who was the calmest guinea pig ever. The kids loved to stroke and fix his long hair. Sometimes, he wore bunches to keep it out of his eyes; and at other times, a fringe (bangs). Whenever he moulted, it was quite shocking to find a big ball of Pippin's hair and a younger-looking, sprightly Pippin.

He always hid in his igloo at feeding time. But one morning, he came out to the edge of the cage. I remember asking him, "Hi Pippin, what ya doing?"

I moved on to feed the rest of the animals. I regret not giving Pippin more attention that morning—he died later that day (from old age). Looking back, I feel he was trying to say goodbye.

Bless you, Pippin.

The only people who seem to have a problem with guinea pigs are some middle-aged women, who have an innate fear of rodents! When you show them that they don't have tails, it usually helps them feel more inclined to get to know the guinea pigs. One Christmas, David pulled a prank on Mrs Claus. He put Terrance the guinea pig in her oven (it was a fake oven—a box with straw, quite comfortable—that she used to produce a burnt pie out of every day to illustrate what an awful cook she was). When Helen reached in for the pie and revealed the guinea pig, the children's mother shrieked with fright and jumped up on the bench. The whole family started crying with laughter. Terrance is very laid back and was delighted with the attention!

EXAMPLE ACTIVITY: FEEDING TIME

If you have a pen that people can throw grass into, it is an amusing activity to watch the guinea pigs eat it like spaghetti. Our guinea pigs have an open-top pen. Children can be very disappointed when the handling sessions are over. I tell them they can spend time in the barn for as long as they want. If they want to go outside and pull some grass to throw into the pen, it will delight the guinea pigs. This leads to hours of entertainment on both sides—another win-win activity! (Guinea pigs can never get enough grass.)

Lessons Our Visitors have Learned from Guinea Pig Handling

- Guinea Pigs help people to develop an understanding and love for animals.
- People learn to be kind to and not afraid of animals.
- They learn not to poke animals' eyes or to put their fingers in their mouths.
- They learn to be calm around animals.
- They learn how to stroke a guinea pig's fur the correct way, and that guinea pigs 'buzz' when they are happy. (People often think they are vibrating out of fear.)

Lessons We Have Learned from Guinea Pig Handling

- Guinea pigs are the top starter animals for engaging people with nature, from the smallest toddlers to great-grandads.
- Guinea pigs are quick to breed! (About 60 to 70 days.) Guinea pig babies are ADORABLE. They are born with a full coat of fur, ready to run around at birth, like tiny little 'mini-me's'. We divided our main housing area into two sections with a plank wall between them. We used to have boys on one side and girls on the other. One day, one of the staff members came running in, yelling, "Guinea pig babies!" How did that happen? The males and females were in their own areas… Someone must have put a male in the female side by mistake at some point. "Oh, how adorable!" Everyone loved the three little fluffy guinea pig babies amongst the eight females. The next day, another staff member announced, "More babies!" Within a week, we had 37 guinea pig babies running around in a mini herd. You could see the shock when people looked into the guinea pig run!
- Luckily it is very easy to re-home guinea pigs to responsible people.
- Males do not live well together in large groups as they can fight for females and bite. Wounds are hard to see through the fur and can get infected. It is best to rear two males together from

weaning; if they have a large enough run and separate igloos, they mostly get along fine (or consider getting them neutered). We have had some male guinea pigs that ended up living solo as they were too dominant and fond of a punch up! If you are managing two guinea pigs during an animal handling activity, make sure they are the same gender. It is a good life lesson, but slightly embarrassing when a male skips a few children to make babies with a female in his line of vision!

- Sexing guinea pigs is virtually impossible. It involves watching a YouTube video and peering carefully, with your glasses on, trying to encourage the penis out. It is a tricky technique and not that pleasant. If we have babies, we mark them up as soon as possible. A little sheep marker spray gets the job done—blue for boys and red for girls. But even so, we get the odd one wrong. The first time we had babies, we asked the vet to do it, and even she made a mistake (which resulted in more babies three months later). As they get older, it is easier to distinguish a male from a female; males have large testicles!

GUINEA PIG CARE AND FACILITIES

Shelter

Ideally, guinea pigs should live in a herd in a large pen. But this is not possible in most situations. Because they are herd animals, you should at least get a pair. A pair of guinea pigs are happy enough in a regular, indoor guinea pig cage. The bigger, the better. They love lots of igloos and tubes to hide in. Guinea pigs use shavings for bedding, which you must replace weekly. They love to sleep in a pile of hay but will be happy in small igloos—either two squished in together or one at a time.

If your guinea pigs live outside, make sure the cage is robust enough to avoid predator attacks and make sure they are not too hot in summer or too cold in winter. Our outdoor enclosure has a durable wire mesh. The mesh is also on the bottom of the run, allowing them to eat grass but not burrow out. It is on wheels, making it easier to move the cage around when necessary. We keep two guinea pigs in a table-top cage for animal handling activities and rotate them every couple of weeks so they all get time to run with the herd.

A friend of mine owns a pair of guinea pigs that came from us. She lets them roam freely in her house. They happily run around and know exactly where to go. They only return to their cage to sleep. She doesn't mind the many poops (they do them all over the place, all the time). When her kitchen door is open, the guinea pigs go outside to eat grass. The cat gets stressed about this and doesn't let them wander too far near the hedge as there are lots of foxes and other predators around, so she herds them back inside!

Food

Guinea pigs squeak with happiness at feeding time. They either need a good, constant supply of grass or specially formulated guinea pig food. The formulated food has the correct amount of vitamin C, which is essential for their wellbeing. They must always have some roughage of either hay or grass. A lot of owners just give the pellets, but this is only half of the diet they need. They enjoy scraps of vegetables, carrots, and greens.

Guinea pigs drink a lot of water, so their drinker needs refilling every couple of days. Be careful on cold nights that the dripper nozzle does not get blocked by an icy drip.

Enrichment

Give them sticks to chew on and pipes or tunnels to hide in.

Health Issues

Clip your guinea pigs' nails every few weeks.

They don't have many health issues, but if you notice a change in behaviour or suspect that your guinea pig has a bite, take it to the vet immediately.

Adopting Guinea Pigs

Always get a pair of guinea pigs. Ideally, get weaned babies, so they grow up being handled.

Older guinea pigs may be nervous but will become comfortable being petted given enough time.

Guinea pigs that have been handled from a young age are incredibly calm.

Training Your Guinea Pigs

Follow the outline for animal handling and make sure the experience is precisely the same for the guinea pigs every day. They will learn what to expect.

Do not give them treats from the hand and only feed them at feeding time in a bowl.

Equipment

- Large cage.
- Igloo/Tunnel.
- Water bottle.
- Shavings.
- Hay.
- Feed.

ANIMAL HANDLING: RABBITS

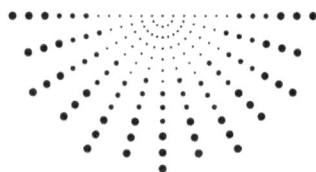

Animal Engagement Rating:

★★

Difficulty of Care Rating:

★★

Expense of Care Rating:

★★★★★

INTRODUCTION

Bunnies are beautiful, and baby bunnies are the most adorable things ever. That's a fact of life!

Our favourite rabbit, Thumper, would happily sit on the petting table for hours and never tried to jump off. We were blessed to have had him with us for many years. One time, it completely slipped my mind to put him away after a school tour. When I came back an hour later, he was still sitting up on the table—waiting for the next group! He could have hopped off the table and hightailed it out the gate if he wanted to...

EXAMPLE ACTIVITY: PET A RABBIT

Time: 15 minutes per group

Rabbits are the most popular of animals for kids. Because they are so fluffy and cute, everyone wants to hold them. That being said, they are not the most suitable pets for children.

Rabbits have incredibly strong back legs for hopping and steel-like claws for grip. If a rabbit hops out of a child's arms, she can leave deep lacerations. If the rabbit struggles and scratches, the child can become afraid and drop or throw the rabbit... This will cause trauma for both for the rest of their lives.

Activity Outline

The only way for children to interact with a rabbit safely is if a staff member (or parent if you're doing the activity at home) places the rabbit on a petting table and monitors the interaction carefully.

Like with the guinea pigs, give the child similar, positive instructions about avoiding the rabbit's mouth and eyes. Bunnies have sharp teeth and rarely bite on purpose, but some develop a nasty habit of biting people. If they do, you cannot use them for interactive activities. Show the child how you gently stroke the rabbit by starting at the head and moving backwards, following the direction of the fur.

The petting table has a low lip to discourage the rabbit from jumping off. Keep your rabbit in a top-opening cage at a suitable level to pick her up out of the cage easily, but with a closure lock that the public cannot open. Train your staff how to pick up the rabbit in a way that avoids scratches.

LESSONS

Lessons Our Visitors have Learned from Rabbit Handling

- People learn that animals have personalities and unique characters.
- People learn to respect animals, especially their teeth and nails.

Lessons We have Learned from Rabbit Handling

- Rabbits scratch, period!
- If a rabbit bites, it is much more severe than a guinea pig. Those teeth are like secateurs, designed to clip off the toughest vegetation.

RABBIT CARE AND FACILITIES

Shelter

Rabbits need a hutch and a run. They love to be outside on the grass.

Food

Rabbits need specific rabbit food with the correct balance of nutrients; this is expensive. They also love grass, fruit and vegetables. They need hay at all times unless there is a lot of grass.

Unless there is a lot of grass available to reduce the cost of their feed, rabbits are quite expensive to keep.

Enrichment

Rabbits like hutches with levels. They also love tunnels and wooden toys to chew on.

Health issues

We had one tragedy with our rabbits. We used to have a pair that lived happily in the chicken enclosure for a few years. But they died suddenly.

We sent the bodies for an autopsy and found out they had caught a disease (Rabbit haemorrhagic disease) from talking to wild rabbits through the chicken wire.

We now have a vaccine that we can give to all our rabbits, so this will never happen again. If you are keeping rabbits outside, contact your vet to find out if they need any vaccinations.

Rabbits are delicate and can get sick quickly. Call your vet immediately if your rabbit refuses to eat or if its behaviour changes.

Rabbits need their nails clipped every four to eight weeks. Their nails are tough, so this is not an easy job. Let your vet do it if you're not sure how to do it yourself.

Adopting a Rabbit

I never recommend that parents get rabbits as pets for their children because:

- Rabbits scratch.
- Rabbits live for 10 to 15 years.

Children have the tendency to get fed up with them quickly, leaving the parents to look after them. This leads to a neglected, unhappy rabbit.

We get tons of calls to re-home rabbits. Choose the ideal pet for a child instead: a guinea pig. They are happy in a smaller run, are more content with being ignored at times and have a much shorter lifespan.

All rabbits are gorgeous, and there are no preferred breeds. I would recommend a small breed (especially if you are adopting your first rabbits). It's best to have a pair of rabbits to keep each other company, and they should be the same gender. Does (girls) can be quieter than bucks (boys).

Beware of mature rabbits in pet shops. People may have returned them because of bad habits or unruly behaviour.

Years ago, before we knew so much about rabbits, we bought a beautiful large rabbit called Sunny for the children from a pet shop. She was a fantastic character and could play football (and was obviously a house pet). Quickly, though, she bit both the children very deeply. We took her back to the pet shop.

The assistant confidently said, "There is nothing wrong with her! Rabbits sometimes nibble or lick you if they smell the salt on your skin. Just get her a salt lick. This is quite normal." She picked up Sunny, who instantly gave her a deep bite on the forearm that started gushing blood.

I don't think she was looking for salt!

Training Your Rabbits

Rabbits are intelligent. You can train them to live in the house and use a litter tray. If you want to learn more, I highly recommend the book, House Rabbit Handbook by Marinell Harriman.

Our adopted rabbits, Ella and Lola—who came to us from fed-up parents after their children had grown up and left home—came with their own three storey apartment. Their bottom floor was the bathroom (where they did all their droppings in a low tray).

Rabbits quickly get used to handling, but you should train your staff how to to pick them up safely without getting scratched.

Equipment

- Food bowl.
- Drop drinker bottle.
- Hay rack.
- Rabbit nail clippers.

FRISKY HERB WALK

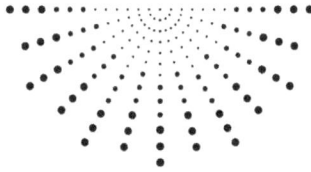

Animal Engagement Rating:

★★★★★

Difficulty of Care Rating:

★★★

Expense of Care Rating:

★★

INTRODUCTION

Frisky, the feral goat, arrived as a gift to the Burren Nature Sanctuary during the first season we were open. She was a few weeks old and had been born on the high Burren hills. Just after she was born, a local farmer startled her herd. Her mother got confused and ran off.

Goat herds can appear and disappear in the blink of an eye. They can travel miles over the hills in a few hours.

The farmer spotted the newborn kid and left her in the field, hoping the mother would return. He went back to check in the evening and found her sitting in the same spot by herself. He then took her home, warmed her up by the fire and gave her a bottle of Frisky (lamb replacement milk) and that became her name. After a week or so of endless bottle-feeding, the farmer got a little fed up and started looking for a home… And so Frisky arrived at the Burren Nature Sanctuary with a bag of powdered milk.

She has since accompanied hundreds of children on school tours, keeping them amused by jumping up on benches to listen to talks about flowers, fauna and karst. Although I try not to have favourites, I LOVE Frisky! She is very tame, and despite her long horns, it is easy to lead her around like a dog by the smallest child.

EXAMPLE ACTIVITY: THE FRISKY HERB WALK

Time: 1 hour

Group size: 1 to 10 adults

Experience outline from Airbnb

We will meet Frisky, the Burren feral goat, at the animal paddocks and learn how a farmer rescued her on the high Burren hills shortly after she was born.

Group members will take turns leading Frisky through the five different Burren habitats: the Hazel and Ash Woodlands, Shattered Limestone Pavement, Wild Orchid Meadow and past the magical Disappearing Lake that drains every 11 to 12 hours.

We will notice which seasonal wild herbs Frisky samples and attempt to identify them.

Get your photo taken with Frisky!

Next, we will learn about the herbal remedies our ancestors used when they lived off the land and the simple recipes they used for common ailments. You can make most of these from garden herbs. Take photos of the Herb Panels with recipes and information on herb properties and traditional remedies.

When we reach the Gypsy Wagon shelter, you'll receive a refreshing home-made elderflower cordial.

Finally, at the end of the walk, you will make your own wild herb tea bag.

You can then relax in the café while enjoying our Burren Food Trail signature dish: a cheese, hazelnut and spinach savoury scone with Burren Gold Wild Garlic & Nettle cheese, and a seaweed salad garnish with home-made relish. You can also pick a freshly baked, home-cooked dessert and enjoy it with tea, coffee or hot chocolate.

Frisky lives with rescued dairy goats, and it is always interesting to see how she, being a feral goat, craves wild plants and herbs. She nibbles on dandelions, wood sorrel, wild thyme, young nettles and wild oregano—everything she needs to boost her immune system.

It can be hard to get people to slow down, to introduce them to plants. Frisky stops the group in its tracks and focuses their attention on grasses or herbs they would normally rush past. All of us must wait for Frisky to devour it and that gives me an excellent opportunity to focus on that plant, name it and discuss it.

EXAMPLE ACTIVITY: FEED THE GOATS

Goats are amiable animals, ideal for adults and children to stroke or feed.

Tell your guests about your goats, what breed they are and what they enjoy doing. Show them that goats have groovy, rectangular pupils. Explain how these allow the goats to see predators in their peripheral vision.

Goats are very gentle and will lick the pellets off your hand. They are ideal for this activity, even for the smallest person. In between animal handling times, we have dispensers ('gumball machines') where you can put a fifty cent in, in exchange for a few pellets (nuts). This little income is a great help for buying feed. And if someone is disappointed that they missed the feeding times, they can purchase a few pellets and feed the goats themselves.

When taking a group of children to feed the goats, the staff member needs to hang on to the bucket assertively! Hold out your hand with your fingers closed and tell the group to show you how they do this. Explain that they should feed an animal with an open, flat hand with all fingers held tightly together. Although goats won't bite, it is good to learn to do it this way. Someday, they might try feeding another animal, like a horse, that could nip them by mistake if they don't hold out their hands correctly.

The staff member should hand out a small amount of feed to everyone. Don't allow little hands in the bucket, or they will try to grab it all. Most of the feed will end up on the floor! Kids will keep on trying to grab food in the excitement, so just smile, lift the bucket and ask them to hold out their hands.

LESSONS

Lessons Our Visitors have Learned from The Frisky Herb Walk

- People learn that animals are sweet and good-natured. They learn that even though animals can't talk, they still communicate and are kind, sentient beings.
- They learn how to hand-feed an animal safely.
- They learn how animals have unique features to adapt to their environments.
- On the Frisky Herb Walk, we get a chance to introduce people to plants that they may have ignorantly walked past all their lives. The next time they see that plant, they will stop to notice it or maybe tell their children about it. The Frisky Herb Walk is the perfect cure for Plant Blindness (see page 114).
- Goats are excellent jumpers. A beautiful Alpine goat that we adopted from a cheese farm, Heidi, was a great favourite with our staff and visitors. She was a huge goat and could jump the fences, however high they were. Whenever she escaped, we would find her in the café, trying to order snacks! Eventually, we made the tough decision to re-home her (in case the health

inspector showed up while she was dining the café). She now lives happily on a smallholding a few miles away and had twin kids this year. She is producing three litres of milk a day!

- We do not cut the horns off our goats :-(but it would be preferable to get 'polled' goats (naturally hornless) for human interactions.
- Friendly goats rarely head-butt humans with their horns. However, their horns are at the eye level of a small child. If the child accidentally fell on the goat (yes, it has happened!), they could get hurt. Un-castrated male goats may run at you and butt you from behind to protect their territory. This is not fun!

GOAT CARE AND FACILITIES

Shelter

Goats need a shelter. A low, rounded tin pig ark bedded with straw is perfect.

Ideally, you should have at least an acre (4046 square metres) of pasture for a maximum of four goats. Fencing is the key to keeping goats. Your fencing should be 5 feet high (1,5 m).

Sheep wire is not ideal. A smaller, grid-like chicken wire is better. The fence needs to be secured at the bottom, too. Our pygmy goats lean hard on the fence until it lifts a little at the bottom and then lie down and slide through under it. For a long time, we couldn't work out how they were getting into the other paddocks until we did a stake-out and discovered their sneaky tactic! Our fence is now clipped securely to a tight bottom tension wire. Goats will quickly work out that they can chew through twine or rope to open a gate, so don't bother using it!

We installed sheep wire, not realising the stumbling-block for our horned goats at the time. Their horns get caught in the fence from time to time. We know the pygmy goats can wiggle back out. However, young goats at different stages of growth can get caught when they reach through the fence as they are not familiar with the size of their own horns. Of course, Google helped us out. We searched 'how to stop your goat getting its horns stuck in the fence'. There were thousands of results. It is obviously a common problem!

At times we have had to use a piece of plastic water pipe, taped to the horns of a young goat with duct tape, protruding out sideways like handlebars. This looks funny, but it is not uncomfortable, and because it is wider than the openings in the sheep wire, it stops them putting their heads through.

If you get your fencing right, goats are a pleasure to keep and just as intelligent and friendly as dogs. They would probably be household pets if it weren't for the fact that they lay a trail of rabbit-like droppings pretty much wherever they go. (I have seen goat diapers being advertised… Google has the answer to everything.)

Food

Different breeds require different methods of feeding.

Our two neutered pygmy goats are continually battling weight issues. They are on restricted diets of hay and little to no pellets most of the time. Goats need a lot of roughage in their diets because they would survive on bushes and leaves in the wild. Milking breeds need more feed. It really takes trial and error to keep the different goats in good body condition—not too fat or too thin.

Younger goats need more food. Some orphaned kid goats find it hard to put on body condition if they have not had colostrum (to set up their digestive systems) as newborns. Our Burren feral (wild) goat, Frisky could typically survive on the roughest mountain tops, but because she was an orphan, she struggles to keep a constant weight. We have the same problem with one of the milking kids that was taken off its mother at birth. They get a full scoop of pellets daily and a vitamin/mineral supplement every six weeks. Get a local goat farmer to come and check the body condition of your goats until you get the hang of it.

Goats love vegetable scraps, young tree branches and leaves; and pretty much anything else you offer them. Don't feed them weird stuff like chocolate as they will eat it, even though it's not good for them.

They have a reputation for eating the laundry from the washing line. I think this is because they naturally like to reach up and browse at things. Goats will eat all the flowers and vegetables in your garden (in a flash) if they escape from their field, and can instantly become unpopular!

Never put their shelter next to the fence. Goats will quickly learn to hop up on their house to jump over the fence!

Key Tip: At Burren Nature Sanctuary, we have one pellet-type feed that is suitable for all the animals. It is a low protein grass pellet. This is to prevent staff and/or visitors from giving fodder to the wrong animal(s). Goats, pigs, horses, alpacas and even rabbits can eat the pellets. The pellet is non-fattening and has a high roughage content, perfect to avoid over-feeding.

Some pet farms hand out bags of food when you pay your entrance fee. Initially, we did this too, but quickly realised on a busy day that the amount of food we were going through was ridiculous; it was not possible to monitor the amount given to each animal, which led us to have welfare concerns.

Enrichment

Goats need to climb. In the wild, they hop across mountain gorges and teeter along cliffs. We have a pile of enormous rocks in the goats' pen, called Goat Mountain. They love playing and sitting on them. They jump up on anything. Our goats even have a slide.

You can search the web for fun ideas for your goats. If you don't have a pile of rocks, you can build play platforms out of timber. Your goats will become depressed if they don't have something to jump up on.

Health Issues

Goats need their feet trimmed regularly. Get the vet or a local farmer to do this the first few times and observe them carefully. Later, you can buy clippers and do it yourself.

The thin wall of the hoof grows under the pad and can trap stones, causing lameness.

Goats need de-worming every three to four months and may need vitamin/mineral supplements.

It took us a long time to discover that Frisky had a copper deficiency and needed a particular supplement that she now gets every six weeks.

Adopting a Goat

Contact the nearest goat dairy operation or cheese farm. The dairy industry does not keep the male kids, and they mostly go to the slaughter at birth. You should be able to take these at no cost.

Male goats must be neutered. Billy goats (un-castrated males) have a pungent smell and cannot be kept for public interaction. If you touch them, the scent will linger after you have washed your clothes and it can be smelled from a mile away. (An excellent signal to other goats, but not so appealing to humans.)

Pygmy goats are lovely animals and are often bought as pets. However, they are boisterous by nature, and although small, they are very, very pushy! They are not suitable as pets for small children. Neutered males will be cheaper to buy than females. We have two neutered males, Mocha and Chino (their mother was a fancy pedigree goat called Latte). Pygmy goats are generally perfect for a visitor centre.

Training Your Goat

Goats can easily be trained to walk on a leash like dogs. Start training your goat when it is still a kid. Find a soft, fabric collar (it must have a plastic snap closure so that it can break if they get caught on a tree), so you don't have to fiddle with buckles. Two fingers should comfortably fit between the collar and your goat's neck when you fit it. Find some tasty treats, and she will quickly agree to follow you on the lead.

Key Tip: Just because your goat agrees to go on a walk with you, it doesn't mean she will go on a walk with someone else! Goats bond with their human friends and unless they like you, they may put the brakes on and bleat like crazy until you go home. Frisky is happy for anyone to lead her if she can see me up ahead. If I am at the back, she will turn around to find me.

Equipment

- Fence.
- Shelter.
- Hayrack.
- Self-filling drinker.

DONKEY PICNIC BASKET

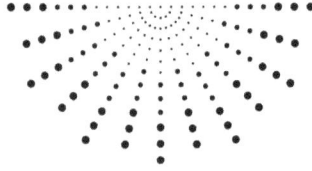

Animal Engagement Rating:

★★★★

Difficulty of Care Rating:

★★★★★

Expense of Care Rating:

★★★★★

INTRODUCTION

Donkeys are hardy, strong desert animals. Until recently, people used them to carry seaweed and turf loads, and to pull carts for transport and haulage in Ireland. They are stubborn by nature and stuck in their habits.

The local farmers used to go to mass by donkey-and-cart. One Sunday, for a laugh, someone switched all the donkeys during the service. The farmers came out chatting with their neighbours and happily jumped on their carts. They shouted, "Hup!" to their donkeys, and all trotted off to the wrong homes… Trying to steer a donkey that doesn't want to change direction usually ends up with the donkey looking back at you but continuing on, regardless.

One of our donkeys, Buzby, does not like the barn. She has never entered it and never will. She is a huge donkey, and no matter how many people push and pull (or tempt her with treats), she will not let one hoof through that barn door. We have given up trying; our farrier must trim her feet in the field! Their stubbornness is part of their charm. They are delightful animals to keep if you have a spare paddock. They love getting attention, are very playful and can spend hours mock-fighting with friends or tearing around the field while carrying an old welly (wellington boot).

Donkeys are an excellent choice to use as therapy animals. They are immensely popular animals, and everyone will love to see and spend time with them.

EXAMPLE ACTIVITY: DONKEY RIDES

Time: 5 - 10 minutes

Number of people per session: 1 per donkey

Most donkeys are happy to let people ride them (train your donkey the same way as you would a horse). However, donkeys can have 'attitude problems'. If he doesn't want to go somewhere, you will need one person leading him and another person encouraging him on. Even then, there's no guarantee he'll move!

We have one riding donkey called Cookie. We used to do donkey rides but always needed two people to persuade him to walk around the paddock: one person in front, leading, and the other person behind, clicking him on... He really couldn't see the point. Remarkably, whenever we had a group of children with special needs, he was always on his best behaviour and didn't need any encouragement. It was very touching.

If you do donkey rides, you can get decorative bridles and saddles with handles (like the seaside donkeys wear). Alternatively, if you're not up for donkey rides, you could do a 'meet, greet and groom a donkey' activity instead. You will, however, need to use a donkey that won't nip or kick. And always have an experienced handler to manage this activity.

EXAMPLE ACTIVITY: DONKEY PICNIC BASKET

Experience outline from Airbnb

We welcome you at the Burren Botany Bubble and give you an introduction to the magical flora of the Burren.

Walk to the Disappearing Lake and continue to the ancient meadow.

Relax in the Gypsy Wagon with a glass of sparkling elderflower juice while Cookie, the rescue donkey, brings your picnic in his willow baskets, which were traditionally used for carrying turf from the bog. Say hello to Cookie and get your photo taken with him. Enjoy your picnic, which consists of our award-winning Burren Food Trail signature dish: a fresh savoury scone with cheddar, spinach and hazelnuts, with Burren Gold Garlic & Nettle cheese; and a seaweed salad. You can also choose a freshly baked dessert from our artisan café: carrot cake, Bakewell tart or a chocolate brownie, all with whipped fresh cream.

Continue on the walks and browse our gift shop before you leave.

This is a popular activity and is suitable for large groups. Get your donkey used to wearing the baskets and make sure he doesn't run off with them. The handler will need more practice than the donkey!

LESSONS

Lessons Our Visitors have Learned from the Donkey Picnic Basket Experience

- People instantly bond with donkeys. (It may be their large, comical ears that make them so attractive.) People get the chance to stroke and talk to a donkey. Most donkeys are happy with the attention. Kids love to ride them, adults love to get a selfie with them!
- People learn about donkeys' affectionate nature and strength that has led them to be servants to humans for thousands of years.

- They learn that donkeys are stubborn and that humans have exploited this trait to use them as beasts of burden; their nature is to not give up, even under the most massive loads.
- They learn that people have disrespected donkeys throughout history. They learn that neglect leads to a donkey having overgrown (long) hooves it cannot walk on. In their natural environment (the desert) donkeys need tough hooves that naturally get worn down by rocky ground, but their hooves don't wear down in soft pastures, hence the need for a farrier.
- They learn about the excellent work donkey sanctuaries are doing all over the world, trying to stick up for this affectionate (but misused) species.

Lessons We Have Learned from Our Donkeys

- People love donkeys! But donkeys have teeth and can bite… You must keep them back from fences by an offset strand of wire if their pasture is close to where the public walk about. People are always trying to feed them. This is not good for them, and if one catches a child's finger between his teeth, the donkey won't notice. But the child will.
- They are especially greedy and quickly learn that they can terrorise tourists into handing over their biscuits and sandwiches.
- They are highly intelligent and will easily learn bad habits. Unless you take the lead and act assertively, they will boss you around.
- Donkeys are noisy… We had to move our donkeys from a field after a neighbour complained that she was up all night with her baby girl, and just when she got her daughter to sleep, our donkeys' hee-hawing woke her!
- Apparently, donkeys can levitate. My friend's donkey didn't like the farrier and levitated (that's the only explanation!) out of a small, top window to escape from the shed before he arrived.

DONKEY CARE AND FACILITIES

Shelter

Donkeys are desert creatures. Unlike horses and ponies, they do not have oil in their coat to protect them from rain. Therefore, they must have a shelter. If they don't have a shelter, they must at least have a rug in the winter.

Your donkeys will use their shelter in hot, sunny weather. But they always appreciate the shade under trees or beside hedges. Make sure there is shade for your donkeys at all times of the day.

You must keep donkeys in pairs. They form strong and lasting friendship groups and typically have a best friend. Never separate a bonded pair; they will grieve for a long time. If there is no other donkey, they may pair with a horse or another animal such as a goat. They need at least an acre (4046 square metres) of field per pair.

Keep their paddocks clear of ragwort. It contains toxic compounds that cause liver damage and can have fatal consequences for equines. Donkeys will not try to eat ragwort while the plant is alive, as it has a bitter taste. The danger, however, comes in when the plant dies. This is when it becomes palatable, and the donkeys will eat it. Hay containing ragwort is particularly dangerous.

Anyone who pulls ragwort must wear gloves as it is toxic and can penetrate the skin. It requires muscles to pull ragwort to remove the plant and roots. So, if you're not too strong, either employ a teenage boy or some sheep, they will happily eat the rosettes of ragwort in the spring (it's not dangerous for them). If the paddock is getting 'stale' with scorched areas of droppings and weeds, it is a good idea to rest it.

Food

As donkeys are not generally worked anymore and survive on slim pickings in their natural desert environment, the trick is not to feed them too much.

In winter, the ideal feed for donkeys is a high-quality straw. It is high in fibre and low in sugar and resembles what they would eat in the wild.

They will also eat hay or haylage, but keep a careful eye on them as they can become overweight quickly.

One sign that a donkey is too fat is a thick upper neck. This is a warning sign. A fat donkey can get laminitis, an excruciatingly painful condition that is difficult to reverse.

Donkeys should not have access to too much grass, especially in spring, when plant sugar levels are high. An excellent method to control your donkeys' body condition is to use strip grazing (this involves 'rationing' the grazing field with an electric fence). You can move the fence 4 feet (1 metre) every day to allow them to graze on another piece of the field when the grass is too plentiful.

We usually contain our donkeys in a smaller paddock during spring and summer and try to keep the grass as short as a golf course's grass from April to June.

Donkeys need barley straw if their paddock is bare.

They can eat any horse ration. However, you should only use it to raise a thin animal's body condition. They love apples and carrots, but do not feed them too many! They also benefit from a donkey mineral lick. Finally, never feed your donkeys grass clippings as it may cause colic.

Enrichment

Bored donkeys will play with anything in the field, but mostly they want to hang out with their donkey friends.

Health Issues

If your donkey's behaviour changes, call your vet immediately.

Some common ailments for donkeys include:

Laminitis involves inflammation of the sensitive layers of tissue inside the hoof. If your donkey is lying down a lot or rocking back on his/her feet, it might be a sign of laminitis.

Colic involves stomach aches or a twisted gut. Symptoms include rolling, looking behind them at their stomachs and sweating.

Overgrown hooves can cause pain and discomfort. Donkeys' feet need to be trimmed every 2 to 3 months.

Donkeys can get small stones in their hooves that cause an infection called 'gravel'. If your donkey is lame and there is no visible injury on the leg, call your farrier to come and examine his/her hooves.

One of our old donkeys, Honey, was a rescue and came to us with a severely deformed foot from years of neglect. After seven years of work, the farrier has encouraged the foot back to a better shape. It still looks a little twisted, but she can walk on it without discomfort. You need a very experienced farrier to work on neglected feet.

De-worm your donkeys every three to four months.

Finally, have your donkey's teeth checked every year.

Adopting a Donkey

If you are considering getting a donkey, contact your local donkey sanctuary.

They will require you to attend a donkey care course and will visit your premises to advise on the suitability of your available space.

Do not buy a jack (un-castrated male), they are cheap because they cause a lot of heartaches... Jacks are naughty and noisy, hee-hawing to the far horizon early in the morning (it can start as soon as 5 AM), waking up the whole county to attract females. You will become a very unpopular neighbour. If you get a young male, have it castrated at about six months old.

Training

If you are interested in using your donkeys for riding activities, call in an expert to train them.

Equipment

- Halter and a rope.
- A well-lit shed or stable for feet trimming. (Your farrier will not want to stand out in the rain.)

- If you are doing rides, I recommend a little saddle with a handle and no stirrups, and a lead rope not attached to the bit.
- If there is no field shelter available, then they must have a winter rug. Small pony winter turnout rugs are perfect.
- Bucket.

A little donkey carried the pregnant Mary to Bethlehem. Some donkeys have the distinctive cross on their backs—supposedly a symbol of their connection with Jesus.

CUDDLE A PET LAMB

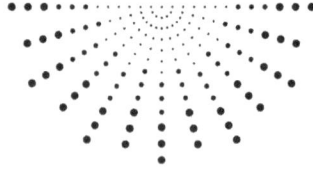

Animal Engagement Rating:

★★★★★

Difficulty of Care Rating:

★★★

Expense of Care Rating:

★★★★

INTRODUCTION

In sheep rearing areas of the world, every spring is lambing season.

Farmers with flocks of sheep will always end up with a few 'pet lambs'. They are orphans who get rejected by their mothers, usually because they are too small and/or too sick to survive. These lambs must be bottle-fed.

Traditionally, these baby lambs were brought into the house and kept in a box beside the fire until they perked up. This still happens on most farms.

Lambs have been smuggled into plate warming ovens for generations!

EXAMPLE ACTIVITY: BOTTLE-FEED AND CUDDLE A PET LAMB

Time: 10 minutes

Lambs need to be bottle-fed two to three times a day with lamb replacement milk.

This activity is suitable for children, but also a wonderfully effective icebreaker for adults from urban areas. Lambs are very friendly, soft and cuddly, and not dangerous in any manner. But they can be surprisingly boisterous. For some reason, Americans in particular are crazy about sheep!

Experience Outline from Airbnb

Guests may arrive at Burren Nature Sanctuary at any time before the experience to explore the trails. (The experience includes an admittance ticket to the sanctuary.)

I will meet you in our lovely artisan café and gift shop and then take you to the barn. Once there, you will meet and cuddle the pet lambs!

Take lots of photos to share your experience with your friends and family.

You can take turns to feed the lambs their bottles. We will give you a bag of animal food, and when the experience is over, you are free to see and feed the other farm animals in the sanctuary (Frank the llama, Hanna and Frisky the goats, etc.).

You may stay as long as you like to explore the trails, the Disappearing Lake, the Botany Exhibit, and to relax in the café.

How to Run this Activity

Keep your lambs in a pen or stable while you prepare the bottles. Get plenty of bottles and divide up the milk, but make sure each lamb gets his quota (either let them out one by one or be prepared to monitor and dive in!)

Sit the visitors down on bales of hay or benches and let out the lambs. Lead them over to the visitors with the bottle and hand it over to your guests. Feeding lambs is an exciting activity. When they're done feeding, people can pick them up and cuddle them.

Show the guests how to hold the lamb securely under the neck and behind the tail. Keep all four legs within the guest's arms. Lambs can try to jump away, so make sure the staff member(s) responsible for the activity receive adequate training.

EXAMPLE ACTIVITY: FEED A SHEEP

Sheep are the second-best choice (after goats) for hand-feeding animals. It is improbable that anyone will get a bite and sheep will gladly take pellets out of people's hands.

They can get boisterous and start bullying and head-butting other sheep (or goats) to get to all the food. (They are very greedy and have no manners.)

Always monitor your sheep to make sure they don't put on too much weight.

LESSONS

Lessons Our Visitors have Learned from the Cuddle a Pet Lamb Experience

- Our visitors get to hold a baby animal, which is guaranteed to make them happy. One sad-looking father was visiting the Burren Nature Sanctuary regularly with his young toddler. He eventually admitted to a staff member that he was depressed and that holding the lambs had really helped lift him out of his despair.
- Visitors learn that baby lambs are feisty and have their own personalities.
- They get to feel how soft and woolly lambs are and marvel at them running around and playing with childish abandon.

Lessons We Have Learned from Our Lambs/Sheep

- Visitors do not want to leave the lamb enclosure area!
- If you are doing this every year, be prepared to keep the sheep forever as part of your experience or hand the lambs back to the farmer after a few months when they're weaned. They will assimilate happily back into the herd. This is an ethical decision as male lambs will end up on the dinner table, eventually. You are doing them a service as pet lambs can do badly if the farmer is too busy and can't be bothered to do all the bottle feeding. A well looked after lamb will make the farmer and lamb happy.
- Sheep are also great as exhibits for nature centres or sanctuaries, but if you receive two or three lambs every year, you will quickly end up with a big herd. We take in orphaned lambs from a farmer who is very kind-hearted to sheep. He has excellent husbandry skills and maintains high welfare standards. He allows us to return the female lambs to him at one year old. After that, they have a lovely life on his farm.

LAMB/SHEEP CARE AND FACILITIES

Shelter

While they're small, you must keep your lambs in a stable at night. They love to be let out during the daytime to run around.

When they're slightly older, they will need a field shelter, like a low pig ark. The round metal variety with one side open is ideal. They need some straw for bedding. Always bring them inside when the weather is bad. Once they're grown up, they will happily live outside.

Food

You will start your lambs off on replacement milk, like Golden Frisky milk powder.

Once weaned, you can start feeding them sheep pellets. It's best to ask your local farmer for advice on the correct variety of feed. They will need a constant supply of grass or hay in the winter. A sheep mineral lick is also advisable.

Health Issues

Sheep need shearing once a year, every spring. Contact your local farmer. He will recommend a shearer in the area. Do your homework and ask for a compassionate shearer as some can be very rough. It is a stressful activity but once 'knocked' (turned upside down), your sheep will not struggle and will feel relieved when the dense fleece is gone.

Sheep need their feet trimmed every six months or if they go lame. The wall of the hoof can grow under the foot and catch little stones, causing lameness.

They need to be de-wormed once a year.

Flystrike: First-year lambs rarely get sheared and are particularly prone to flystrike. A fly will lay eggs in the wool, and maggots will eat the lamb's flesh once they hatch. You may notice a patch of low wool, wool falling off the sheep, or little bits of wool around the field. By the time you see a bald spot, the infestation is already severe.

Preventative measures include dipping sheep in chemicals. However, it is usually unnecessary if you monitor your sheep. You treat flystrike by pouring a solution on the sheep's wool to kill the maggots. Call your vet, or ask a local farmer for help if you notice flystrike on any of your sheep. This condition can become severe, cause extreme suffering and can be fatal if left untreated.

Sheep can get several diseases. The only other one we have experience with is orf (also called scabby mouth). It is a viral infection that manifests as scabs around the mouth. It is highly infectious to other sheep (and goats) and can be transmitted to humans.

If you suspect one or more of your sheep have contracted orf, isolate them immediately, and treat and disinfect any areas they have been in. This is a self-limiting disease, meaning it will clear up on its own without treatment within three to four weeks. The virus can 'hang around' and cause outbreaks in other animals.

Adopting a Pet Lamb

Ask your local farmer or search for pet lambs on farm trading websites. You will find many pet lambs being advertised as it is expensive to rear them, and farmers don't want to devote the time to them. Farmers will put down some of their orphaned lambs if they cannot find foster homes for them.

Adopting a Pedigree or Rare Breed of Sheep

It is vital to do research on the breed of sheep you want to purchase, especially if it is a rare breed. Ancient breeds, such as Jacob sheep, are lovely animals. They are straightforward to keep and don't need human assistance during lambing. They are, however, expert escape artists. The whole herd will jump a high fence in the blink of an eye. Once they've learned how to break out and start making a habit of it, it becomes impossible to re-train them... The behaviour is irreversible!

A friend of mine wanted Jacob sheep to breed. I gave her my worst behaved ewes for free! She put them in her yard to check them over before letting them in the field. The wall of the yard was 8 feet (2,4 m) high. They jumped out and lived in an area of commonage for months. It took a large group of people to capture them!

When our new neighbours moved in, the lady said to me one day, "It's lovely seeing the wild goats come down from the hills into our garden." The 'wild goats' were, in fact, our Jacob sheep that had crossed the busy road to feast on her flowers.

A few years ago, the Department of Agriculture produced a brochure called 'Staying with Sheep' in an attempt to encourage more people to farm with sheep. Upon seeing the brochure, my friend's young son commented that they should have called it 'Convincing Your Sheep to Stay with You'!

Training Your Lamb/Sheep

Sheep don't need training. They are shy but friendly animals. If handled from a young age, they will be effortless to manage.

They quickly learn about feeding time!

Equipment

- Hoof trimmers.

COLLECTING EGGS AND FEEDING POULTRY

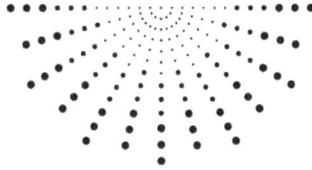

Animal Engagement Rating:

★★★★★

Difficulty of Care Rating:

★★★★★

Expense of Care Rating:

★★★★★

INTRODUCTION

This is a popular activity if you have poultry. Hens and ducks lay eggs every day during the summer, mostly in the same spot.

Hens and ducks can, for the most part, live together in peace. Sometimes hens can bully ducks; and if you have a rooster, he may be mean to the ducks.

If you have ducks (so adorable!), you should have a pond for them. We have a smallish paddling pool that we empty every day, but a larger pond would be ideal. Ducks will get depressed if they can't swim every day.

My friend had chickens and ducks. After a fox attack, only the rooster and one duck survived. The duo trailed off to the lake every day, where the duck would swim around, and the rooster would wait patiently on the bank. They always went home together at sunset.

At my grandparents' farm, we used to have a chicken called Henny Penny. She had no problem with being dressed up in a headscarf and being pushed around in a pram… Both children and hen seemed very keen on the activity!

EXAMPLE ACTIVITY: COLLECT EGGS AND FEED THE CHICKENS AND DUCKS

Put on wellies!

Have hand washing facilities near the pen.

Hand out baskets to the children and enter the poultry area. Check the nest boxes and if there is a broody hen lift her up to collect the eggs. You can also catch a quiet chicken or duck for children to hold.

Remember to wash your hands when done.

Key Tip: put feed into a small hutch and then retrieve the chicken from above. Chasing chickens may be fun for children, but not for chickens. A docile hen can get used to being cuddled and can easily be tamed. Chickens, despite their reputation, are exceptionally intelligent.

LESSONS

Lessons Our Visitors have Learned from Collecting Eggs

- Visitors learn where eggs come from and to connect what they eat with the animal.
- They learn that chickens love running around, scratching and foraging outside and that this is what 'free range' means. If appropriate, develop this conversation into a discussion about 'cheap chicken' and 'factory chicken', and why it is a deeply flawed system. During such discussions, our visitors learn why it is essential to buy meat from reputable sources with good welfare policies in place.

- Our visitors come to understand that chickens are sentient beings with unique characters and temperaments.
- They learn that chickens are nervous creatures that fear predators. They also learn that foxes love a chicken dinner.
- Finally, visitors learn that chickens can be loving companions once they trust their human keepers.

Lessons We have Learned about Poultry

- Be careful if you have any poultry that peck, for example, a rooster. We have a duck that pecks me on the ankle every day to hurry me along to the bag of oats!
- There may not always be enough eggs to collect for each child in a group. It can lead to disappointment and tantrums. Oh well, you can't win them all...
- Chicken poop is unhygienic and can be a frustration in wet weather as it will get on shoes and/or boots. As with all our animal activities, we always advise people to wash their hands after feeding and handling the chickens.

POULTRY CARE AND FACILITIES

Shelter

A hen house with accessible nesting boxes is ideal. The poultry enclosure must be safe from predators. We have a 6-foot (1,8 m) high chicken wire fence. The fence is dug into the ground at least 6 inches (15 cm). We have an electric wire on the top and bottom.

Despite our security measures, we have had a fox attack. He managed to jump onto a corner post, balanced on one foot, and hopped in. Foxes will check your chicken enclosure every night without fail.

You should commit to putting your hens in every evening and checking that their pen is secure. If a fox attacks, it might attempt to kill all the hens and remove them to a food store to eat at a later stage. They usually don't have enough time to carry all hens away though, and that's why people think they kill them for fun. Usually, they take fright thinking they have been discovered before they have removed all their haul.

We have also had a pine marten attack. Pine martens will kill all the hens in an enclosure. Unlike foxes, they cannot carry hen carcasses away as they are too heavy for them. Also unlike foxes, pine martens have been known to kill for fun. You can identify their kills by two puncture marks in a hen's neck or by decapitated hens.

Dead hens are a sad sight to see, but it is also a learning experience. It helps to keep in mind that foxes and pine martens see chickens as part of the food chain and that they may have young to feed. Predators are a natural and essential part of life.

Food

Buy good quality chicken pellets. Chickens will eat plain oats but will need a balanced feed containing grit to produce eggs.

Feeding chickens can be an intimidating activity for children if you go into the pen as the chickens will rush to them and may peck at their feet.

At the Burren Nature Sanctuary, we have a pellet dispenser outside the chicken pen. We find that children enjoy taking pellets from the dispenser and throwing it to the chickens through the fence.

Health Issues

Chickens have a wide range of health issues; most can be avoided by providing them with hygienic living conditions. Keep your hen house and poultry area clean. If illness breaks out in the flock, the whole area will have to be disinfected.

Adopting chickens

For information on adopting and training chickens, please refer to the previous section on chicks.

Equipment

- A fox-proof hen pen (enclosure) and henhouse.

SELFIE WITH AN ALPACA OR LLAMA

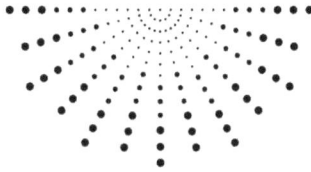

Animal Engagement Rating:

★★★★★

Difficulty of Care Rating:

★★★★★

Expense of Care Rating:

★★★

INTRODUCTION

We were unsure whether to adopt Frank the llama and Jazz the alpaca. Both were looking for homes; Frank was an orphaned llama that had been hand-reared, and Jazz was an unwanted senior male from a commercial herd. In the end, we decided that although they do not represent the local fauna, we were happy to welcome them as 'blow-ins'. A 'blow-in' is a piece of seaweed that has blown into the land. Locals welcome it as a fertiliser for the fields. Before Kinvara became a tourist attraction, it was a market town for the surrounding area. Traditionally, 'blow-ins' were welcomed with open arms as they represented good fortune and brought wealth with them. Kinvara residents still welcome outsiders with gladness and kindness.

Llamas are the most eccentric animals you will ever meet. Individuals ooze with character, and they are stunningly beautiful. They have the coolest stride when they walk and, when they want to run, it looks like they're skipping and bouncing all over the place!

Alpacas have a sweeter, shyer nature than llamas. They are not naturally friendly. As herd animals, they are nervous of predators—especially coyotes (their natural predator). Farmers will often let an alpaca graze with sheep as a guard from foxes. The alpaca will run after the fox and kneel on it to break its back. If they see a dog, they will sound an alarm (and even attack it if given a chance).

EXAMPLE ACTIVITY: SELFIE WITH AN ALPACA OR LLAMA

This activity is easy: get your phone out and take a photo while posing with an alpaca or llama. Llama selfies go viral!

Frank, our llama, was orphaned and consequently reared with a bottle. He is used to human interaction and is happy to pose for as many selfies as needed. He is quite convinced he is handsome. Besides, everyone tells him so every day. Now that he is older, though, he is more discerning and randomly decides who he is prepared to pose with.

Alpacas and llamas really don't need to do anything. People just love looking at them. They are so exotic and comical-looking and make great photos. Alpacas are shy, mild-mannered animals. Their main form of protesting is sitting down. If they don't want to go somewhere or do something, they plump down.

FUN FACTS:

- Alpacas sit down to mate!
- They can spit, but I have never seen our Alpacas do it.
- If you want to feed an alpaca, extend your hand into the paddock, and try to not look it in the eye (especially if you know he is shy). He will gently pick the pellets off your hand.

- Alpacas don't really like being petted. Frank the llama, on the other hand, was reared with a bottle and is used to being rubbed.

LESSONS

Lessons Our Visitors have Learned from Jazz and Frank

- Visitors learn that animals are amazing and funny!
- They learn that animals have adapted their physical appearances and behaviours to survive in different environments. Alpacas and llamas have long necks so they can see predators from a distance and thick fur to help protect against animal bites. Alpacas have padded feet so as not to damage the scarce vegetation in the Andes in Peru (where they come from).
- People learn to respect personal boundaries. If Frank does not like what you are doing, he won't hesitate to tell you by pulling his ears back and gathering some lovely, gooey green spit!

Lessons We Have Learned from Jazz and Frank

- Do not house two males in adjoining paddocks... They will fight. When Frank (the llama) arrived, he and Jazz (the alpaca) had a major show-down over the fence. Frank carries a split in his ear as a reminder. They still get angry when they see each other in the distance!
- I only know of one incident where Frank spat at a human. I saw a video that went viral; it was of someone skipping along a fence with an alpaca. It was so funny—I had to try it with Frank the llama... He joined in, and it was hilarious. The next day, Helen would host a group of international tourists on The Fairy Pig Walk. So, I said to her, "You have got to do the skipping with Frank. They'll love it! Just make the group skip along the fence and Frank will join in. He loves it." Of course, Helen told everyone to skip along the fence the following day. Frank became upset, ran to Helen and spat, covering her in a stinking, green sludge—from her neck to her knees. The group just stared in shocked silence, and this was before the walk had even started.

Later, Helen told me, "You were wrong… Frank does NOT love skipping!" Oops…

- Because Frank was reared by humans, he now thinks humans are other llamas and decides who he likes (or doesn't like). This behaviour is common in hand-reared male llamas. In the wild, once they reach puberty, males start 'negotiating' their position within the herd. Basically, they establish their rank by picking small fights. As a baby, Frank was happy to cuddle with everyone because (from his point of view) humans were higher ranking llamas than him. But now that he's grown up and thinks he's the boss, he maintains the 'right' to pull back his ears and gather a ball of spit whenever he feels the need to assert his authority.

- Frank loves people a little too much. He has been known to try and 'cuddle' people, especially women, who go into his paddock and are not paying attention. Being stalked by a llama and suddenly feeling two hooves on your shoulders could leave you with lasting PTSD! Frank unashamedly did this to Helen on a busy day in front of an audience. She bent over to pick up a guinea pig from the outdoor run, and the rest is history… To add to her embarrassment, visitors were upset with her for swearing at Frank.

- Alpaca shearing is very stressful. (See the Alpaca/Llama Care and Facilities section on the next page.)

ALPACA/LLAMA CARE AND FACILITIES

Shelter

Alpacas and Llamas need an acre (4046 square metres) of land per pair. They are very clean animals; they poop in one spot. Their poop (which doesn't need composting as they have three stomachs to do all the hard work) makes for a valuable fertiliser. I have seen a small bag of alpaca poop, the size of a bag of coffee, selling for a whopping €10.

You can fence them with low sheep wire (4 feet/1 metre). They don't seem to be interested in jumping fences. A llama shelter should be tall enough for a llama to fit into when fully grown (they can grow up to 59 feet [1,8 m]).

We have never managed to persuade any of our alpacas to use their shelter. We tried one with a wooden floor and another one with a grass floor. The alpacas couldn't care less about all the trouble we went through for them, and no matter how hard we tried, they would not set foot in their shelter. They continued to stay out in the pouring rain. We've learned that they sit down and keep the spot underneath them dry when it rains. They seem to be okay outside during the whole year.

Alpacas are well adapted to handle cold weather due to their natural environment, but even so, it is not natural for them to be outside in constant, heavy rain. Our alpacas are elderly, and it's painful to see them miserable and wet. We now house them between October and February every year and let them out for exercise during the day. You can also consider getting waterproof coats for your alpacas during winter months.

Food

Alpacas will eat sheep pellets and hay. They can eat high fibre, low protein horse pellets, but be careful not to give them a high protein pellet.

Llamas seem to only eat grass or hay and maintain good body conditions. Our llama, Frank, doesn't like any grains or pellets. I am not sure if this is a trait of the breed. He eats a lot of hay and likes to chew pretty much all day long. When people try to feed him pellets, he takes it from their hands to be polite and then spits them out on the ground.

Enrichment

I have heard that both alpacas and llamas enjoy toys, but ours are happy just nosing around their environment. The most important thing for them is to have enough space.

Health Issues

Woolly llamas need a yearly shear. Frank is a hairy Llama, so he does not need shearing. If you adopt a llama and you are unsure whether he is hairy or woolly, contact your vet to examine him and advise you.

Alpacas always need a yearly shear. They do not shed and can develop health problems when left unshorn. Alpaca shearing is difficult and requires skill, as alpacas absolutely hate the process. Our local alpaca shearer is, thankfully, a master in his craft.

Jazz is at least 17 years old and has been sheared many times in his life. Still, every time is a traumatic experience. It always involves him crying, fighting and needing to be pinned down. The shearer holds him down securely and will sometimes tie a rope around one of Jazz's feet.

Shearing is stressful for both alpacas and their owners. It is not an event for public viewing. Our shearer does the job as calmly and quickly as possible and is always careful to not harm the alpacas with the shearing blades. He also trims their feet and teeth, and he gives them a copper supplement which they need for their digestion.

Alpaca teeth need to be trimmed because they continue to grow. In their natural environment, they would eat tough plant material, and this would wear down their teeth (like a donkey's hooves would wear down naturally in desert plains). Hay and feed are not tough enough to wear down their teeth, hence the need for trimming. Teeth trimming on its own is a speciality and requires a qualified person to work with a particular machine, but our shearer has many years of experience and trims our alpacas' teeth with special pliers when he comes to shear them. He has found that they are terrified of the sound of the machine and nipping them off with pliers is quick and less traumatic.

Both alpacas and llamas need to be de-wormed yearly. It may be hard to give them their de-worming medicine orally as they can refuse to take it. If that is the case, consider an injection at shearing time.

Adopting an Alpaca or Llama

Alpacas and Llamas are herd animals and at their happiest in pairs. In our experience, llamas seem to have less of a herd instinct than alpacas. Alpacas seem to thrive in groups and will be just as happy amongst a sheep herd. I suggest you adopt a pair of females or a male-female pair. Do not choose two males to live together. Two neutered males who grew up (or will grow up, if you adopt crias) together might work, but we have never tried this after the episode with Frank and Jazz. We have had Frank neutered, but he still has a considerable problem with Jazz.

Alpacas and llamas are beautiful, easy animals to keep and with a little experience, can be managed easily. It's easy to move them between fields or into the barn. They don't usually run off in random directions like sheep. They don't try to break out of their paddocks and do all their droppings in the same spot. They are a joy to own!

Training Your Alpacas/Llamas

Alpacas and llamas enjoy mental stimulation and can be trained. Depending on the individual, it may require perseverance on your side. I bought a lovely little halter for Frank (the llama) and got it on him without trouble. However, when I tried encouraging him to walk with it, he exploded into a fit of somersaults, bucking and rearing. He fell over and blindly ran into the fence.

I did not try it again! Because of my experience, I believe it's best to train them from a very young age.

Equipment

- Shelter.
- Hayrack.

MILK A COW

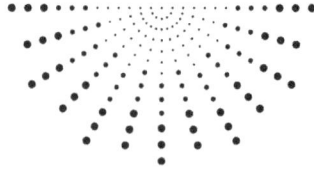

```
Animal Engagement Rating:

★★★★

Difficulty of Care Rating:

★★★★★

Expense of Care Rating:

★★★★★
```

INTRODUCTION

Because of climate change and global warming, we decided not to breed our cows any longer, so we don't milk cows as an activity at the Burren Nature Sanctuary.

As all vegans know, dairy is an emotive topic… Most milking cows' calves are removed from them, and instead of drinking their mothers' milk, they receive substitute milk via bucket feeding. Cows have strong maternal instincts. Losing their calves due to human intervention is traumatic, and they will mourn the loss of their calves.

If you have a docile cow who doesn't mind being separated from her calf for a while, you can run a cow milking activity. Just make sure her calf is in a nearby stall.

It is an engaging activity and will always be a popular attraction for your nature centre.

We have always been an organic farm and had a small suckler herd. Suckler herds are kept for beef production. (as opposed to milk production). Their calves stay with them until they are ready to be sold.

We do not breed the cows anymore because we are now a nature centre. Our three cows are retired, and their job is to decorate the fields and manage the grass in the paddocks. They are Buckle and Beltie the Scottish Galloway cows, and Petal the Scottish Highland cow.

Our three girls are not traditional Irish breeds, but they are well-suited for the environment and can stay out all winter without being housed. The sanctuary's soil drains well and is suitable for overwintering cattle. They came to us as weanlings (just weaned from their mothers).

Despite their lovely personalities, they are not tame. At most, they will come up to us to eat food from a bucket. Their somewhat feral existence can be troublesome for our forbearing vet, who is especially cautious of Petal's enormous horns!

One of our Scottish Galloways, Buckle, is not impressed with our 'no breeding' policy and has, for the last three years, knocked a gap in the wall and gone in search of a neighbour's bulls. It is quite embarrassing. The farmer must gather his herd, rebuild his wall, and bring Buckle home. Despite all the 'drama', our neighbours have been most kind and long-suffering. This year, Buckle chose a pedigree, prize-winning Limousin bull called Masonbrook Legend. She surprised us with an adorable, big baby boy as I was writing this book. Apologies to the climate… Our new baby has the same white 'belt' as his mother. This is the first time she had given birth to a calf with a 'belt'. Check out our Instagram page for pictures:

https://www.instagram.com/burrennaturesanctuary/

Her sister, Beltie, has no idea where Buckle goes every year and has failed to grasp the need to rear children. She alerts us by mooing for Buckle when she disappears.

EXAMPLE ACTIVITY: MILK A COW OR GOAT

Time: 30 minutes

We used to have the most beautiful jersey cow, Angel. She had been a 'house cow' all her life. In Ireland, 'house cows' live on non-dairy farms to produce milk for the family.

Angel was so calm, you could just walk up to her in the field with a bucket and milk her. One year, we selected a Jersey bull from the AI man (artificial insemination) to breed her. The AI man arrived in a van with his frozen semen straws and I took him out to Angel in the field.

"You need to put her in the crush." He said.

"Oh, we don't have a crush. I can put a head collar on her if you want, but she won't mind."

He shoved his hand under her tail and manoeuvred the straw into the right position.

"Well, that's a first. I have never inseminated a cow standing in the middle of a field."

She really was an angel.

Activity Outline

You will need a stall like a cow crush in your barn. Ideally, the stall should be raised 2 feet off the ground. It should have a ramp to lead the cow (or goat) up. Goats are easier to manage than cows during milking.

If you use a cow for this activity, make sure she does not kick. Some cows will always try to kick, and others will settle once they get used to it. Place the milking stool to the front of the udder and demonstrate how to milk the cow. This is an extremely popular activity, but it's not simple to orchestrate. During milking activities, you will have to separate her from her calf. She may or may not agree to this. Most likely, she will bellow loudly in protest.

Once her calf is weaned, you must milk her out twice a day (she can get mastitis if you don't). If you want your cow to produce milk continually, she must have a calf every year.

You can also let people feed cows as an additional activity at your nature centre. Cows are generally not feisty, but they have immensely powerful heads and might bang into someone like a wrecking ball by mistake. It is best to exercise caution.

LESSONS

Lessons Your Visitors can Learn from Milking Your Cows

- Visitors learn where milk and cheese come from.
- They connect the product with the animal, and it creates the possibility to talk about animal welfare.

Lessons We have Learned from Running a Cow Milking Activity

- This is an activity that can be done successfully if you have the experience and the necessary infrastructure and equipment in place.
- As an alternative, you can consider using a fake cow with a rubber udder filled with water. It is undoubtedly much easier to care of and manage!

COW CARE AND FACILITIES

Shelter

Cows are herd animals and need a big range. The minimum outdoor space is 1 acre (4046 square metres) per cow. Since they are herd animals, you need to get at least two. Depending on the breed, your cows may need to be housed in a stable during winter months.

Do thorough research about your chosen breed to make sure you can care for them properly.

Food

They will need a continuous supply of hay, haylage or silage as soon as the grass fails in autumn. Either cut your own or buy large, round bales. Expect to feed your cows six months of the year.

Enrichment

Cows do not need additional enrichment if they have large fields to roam around.

Health Issues

Cows can suffer from various ailments. If you know your cows well, you will be able to tell if something is wrong with them when their behaviours change.

If you notice such changes, call a vet who specialises in bovine (cattle-type animals) care for help.

Testing for bovine TB (Tuberculosis) and Brucellosis must be done every year by law in Ireland. Both are contagious bacterial diseases. This testing involves an injection, and the cows know precisely when the vet is due to come. They usually do a disappearing act just before he arrives!

Adopting Cows

My suggestion is that you arrange a meeting with a local farmer for advice.

Training Your Cows

We have never had the desire to train cows, but it can be done!

Equipment

- Cow crush.
- A round feeder for bales.

GROOM A PONY

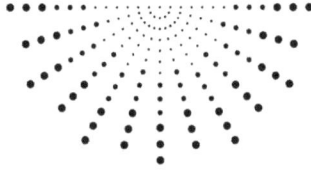

Animal Engagement Rating:

★★★★

Difficulty of Care Rating:

★★★★★

Expense of Care Rating:

★★★★★

INTRODUCTION

All kids love ponies, but not all ponies love kids. Managing equine interactions requires experience.

We have two miniature ponies, Molly and Vincent. Molly is a rescue pony. Two years ago, a miniature stallion yearling named Tiny (original name, I know!) needed temporary lodging, and we gladly welcomed him. Months after he had left, Molly surprised us with the sweetest little foal. Vincent was smaller than a Labrador retriever and looked like a soft toy when he was born.

Both Molly and Vincent are a beautiful strawberry roan colour with white blanket markings. They are lovely to look at but are fenced a good 3,2 feet (1 m) away from people's fingers.

We also have Toffee and Bailey, two palomino, Haflinger x Clydesdale cobs "Palomino" refers to a genetic colour in horses, consisting of a golden-brown coat and a white mane and tail. They were gifted to my girls as yearlings when their uncle emigrated to Australia. They are trained to ride but spend most of their time 'decorating the fields'. Everyone loves to take photos of them.

A little girl once asked me, "Where are the Barbie horses?"

I think it is safe to say everyone loves horses. But equine interactions, especially riding, are high-risk activities. For that reason, we do not offer it. Riding a horse is a wonderful pastime that instantly connects you with Mother Nature and the Earth. I highly recommend that you book riding lessons with your local stables or even a family trek. Trekking horses are typically docile enough for beginners to hop up on and plod along.

EXAMPLE ACTIVITY: GROOM A PONY

If you have a small, quiet old pony who is happy to stand tied to a wall or post, and you are 100% confident he will not kick, you can let kids brush and clean him.

You can show them how to brush out the tail and how to plait the mane, then let them do the plaits and roll them up with plaiting bands. Pony-mad children will love to learn about the parts of the horse, parts of the saddle and bridle, and the names of the grooming equipment. They have an insatiable desire for pony knowledge.

LESSONS

Lessons We have Learned from Our Ponies

- Ponies can tread on (and potentially break) your toes. You won't forget that experience in a hurry! Show the children how to approach the pony and how to stand to the side, while being aware that the pony might move.
- If your pony is docile and willing, you can make a small timber

pen for him to stand in. This will prevent him from moving sideways and squashing or treading on a child's delicate feet.

- Children always want to ride on the ponies. You can decide if it is safe to allow the children to sit on a pony while you hold them. But doing this increases the safety risk of the activity, so you need to think about whether you are willing to take the chance (and whether you are insured).
- Ponies love being brushed, and children love interacting with ponies—so this is a win-win activity.

PONY CARE AND FACILITIES

Shelter

Owning ponies is a big subject and requires a book in itself. If you are not experienced, make friends with someone who is. They must have a paddock (at least 1 acre per pony) and a shelter or stable.

Food

Their diet must be monitored carefully, small ponies can get dangerous illnesses from overfeeding. We mostly feed our ponies grass, with supplementary haylage over the winter (hay that has not dried completely) Get expert advice if you are not experienced with equines.

Enrichment

Ponies want space to range around.

Health Issues

Ponies need their hooves trimmed every six to twelve weeks. They need shoes if they're being ridden on tarmac. This is an expensive commitment and shoes cannot be neglected; they need to be changed every six to eight weeks.

Ponies kept outside are generally healthy. As with donkeys, call your vet if you notice behavioural changes.

Laminitis is a severe risk for fat ponies, especially during spring. Grasses that grow during springtime have a high level of fructose, which can cause laminitis in susceptible equines.

As with donkeys, small stones can get lodged in the underside of ponies' hooves. It can cause lameness and infection. If your pony is lame, call the farrier. If he can't solve the problem, call the vet.

Horses/ponies must be de-wormed every three to four months.

Finally, they will appreciate a salt lick.

Adopting a Pony

With horses and ponies, younger is not always better. Horses can live well into their twenties (and even thirties). If you're adopting a horse or pony for the first time, it is better to get an older animal.

Only buy or borrow a pony that a friend has recommended. An old children's pony is a good sign; he has probably been well looked after for so long because of a good temperament.

Take the pony for a trial of at least a week if you are serious about purchasing. Ask advice from someone who really knows about horses (you will find many people keen to give advice, but there are only a handful who actually know what they are talking about). Choose a gelding (a castrated male) as they usually have the best temperaments. Mares can be irritable (more often than not). They pull their ears back and swish their tales to show how annoyed they are. Never purchase a stallion (an uncastrated male).

Horses and ponies are herd animals and need companions. A companion doesn't necessarily have to be another horse. They are quite comfortable with the company of cows, sheep, goats and even dogs or cats.

Training Your Pony

Do not attempt to train your pony unless you are an experienced horse person. A good habit takes years to instil, a bad habit can be picked up in a minute and is often irreversible! Get help from a friend with lots of experience or hire a professional trainer.

Equipment

Note: this is not an exhaustive list. It is the bare minimum you will need if you buy a pony. They are expensive to keep and require a significant commitment on your part.

- Headcollar.
- Bridle for better control.
- Rope.
- Bucket.
- Winter rug (if you have no shelter).
- Grooming kit and carrier.
- Body brushes, dandy brush, curry comb, hoof oil and brush, plaiting bands, sponges.

WATCH THE FISH

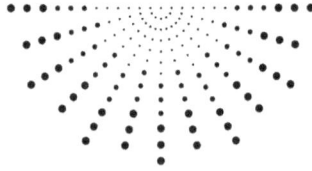

Animal Engagement Rating:

★★

Difficulty of Care Rating:

★

Expense of Care Rating:

★

INTRODUCTION

There is not much to say about goldfish… They are just great. Someone donated a pair of goldfish to the sanctuary, and they now live happily in the water feature, in the Burren Botany Bubble. One day the water feature sprung a leak and drained; we were so devastated for the fish… But when we refilled it, to our delight, our precious goldfish were alive and well. They must have found a puddle in a cave to hang out in.

EXAMPLE ACTIVITY: CHILL WITH THE GOLDFISH

People love visiting our goldfish. Watching them swim around is, perhaps surprisingly, meditative. I met a dad and his toddler in the Botany Bubble one day. The dad told me, "Every time we come to Burren Nature Sanctuary, we have to visit the fish before we leave, or there is a tantrum!"

Lessons Our Visitors have Learned from Our Goldfish

- Visitors learn that even little goldfish have their own personalities and are mesmerising to watch.
- The fish are brought into the conversation when we tell visitors that the Burren was a tropical sea three-hundred-and-fifty million (350,000,000) years ago, and that ALL the stones in the hills are made up of tropical fish bones.

Lessons We have Learned from Our Goldfish

- A goldfish's environment can influence its growth. Our goldfish have grown quite large as they have a lot of space to swim in, and the pond replicates their natural habitat.
- Goldfish in captivity need time to digest their food, so they don't need to be fed every day.

GOLDFISH CARE AND FACILITIES

Shelter

A goldfish can live in a bowl because of its hardiness. But if you can, get a proper glass tank to house your fish. Better yet, house them in a pond. Your tank should be rectangular and at least 20 gallons per fish, plus 10 gallons per additional fish. In a worst-case scenario, you can house two goldfish in a 10-gallon tank. Never go smaller. Goldfish create a lot of waste, so you will need a strong filter for the tank. You should clean their tank at least every week (you can stretch this to once every two weeks at most). Your tank will have to be cycled. This means that you will give beneficial bacteria time to grow. These bacteria break down waste and minimise ammonia levels (which are harmful to your fish). Do research about the best cycling methods.

Before the pet shop industry became popular, my husband got a goldfish at a fair in the 70s. Because fish tanks were not readily available as they are today, his father made a custom tank with glass sweet jars, with tubes connecting them. Of course, they used to love watching the fish swimming through the pipes to their different rooms. Good job, Mick! RIP.

Food

Goldfish, especially fancy goldfish, do not need feeding every day. They can overfeed easily and develop digestive problems. You can give them food every other day. You can feed them lightly steamed vegetables like carrots and peas besides their normal fish flakes/pellets.

Enrichment

They appreciate enrichment in their tank or pond, even old plastic toys, like SpongeBob figurines, that they can hide behind or swim through.

Health Issues

Common issues you may encounter with your goldfish are swim bladder disease and white spot disease (also called ich).

Adopting Goldfish

Visit local buy-and-sell websites to see if anyone is selling fry (baby goldfish) or wants to re-home their own goldfish. The benefit of taking in someone else's fish is that you won't have to spend extra money on a tank and equipment (unless what you get is inadequate). Always get more than one goldfish as they are social creatures.

We would like to keep tropical saltwater fish to illustrate the amazing fact that the Burren used to be a tropical sea. Keeping tropical fish requires a heated tank and even more attentive care than with freshwater fish. Freshwater tropical fish are easier to care for but to keep the 'Finding Nemo' varieties—that children love—a saltwater tank is required. This is a specialist area, as salinity must be carefully maintained, and it can quickly go wrong if you don't have an experienced person in charge. They are also an expensive item.

In comparison, goldfish are very user friendly!

Training

Yes, you can train goldfish! They are intelligent and enjoy mental stimulation.

Equipment

- A rectangular tank of at least 20 gallons.
- Tank filter.
- Tank plants and accessories.
- Fish food.

MAGICAL STORY
THE MAGIC HORSE

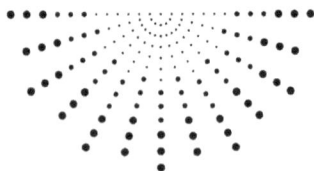

When I was nineteen, I travelled to Ireland during my summer break from University and my life took an unexpected turn.

A year later, I sat huddled next to the gloomy grey ash of an open fireplace. There were no more briquettes to keep the fire going. I would have loved a cup of tea, but the one-gallon churn for our water was empty. The well to refill it was half a mile away. An injured black terrier sat at my feet.

The baby was asleep in her crib, wrapped up warmly. When she was born, they hung her up under her arms for me to see. She surprised me. I'm not sure what I expected, but I certainly didn't expect a baby to be a 'person'. And I never expected those big, blue eyes to lock on to mine as they dangled her in front of me. Back when she was only a heated topic of conversation in my womb, I had vowed to look after her. But I hadn't considered that I would love her wholly and instantly.

In January, when the baby was six months old, I went to muck out the horses. There were two five-year-old cobs in for breaking: a bay and a chestnut. They belonged to an old farmer who had swapped his carthorse for them at the Ballinasloe Fair. We had already found a buyer for the bay. The chestnut was a runaway.

Two neighbours showed up and took the small chestnut horse (only 15.2 hh) out into the front paddock. They put a heavy pole on two metal barrels, clipped a lunge rope on the horse and tried making him go in a circle. The chestnut was unimpressed and kept pulling backwards and getting away. They had to chase him into a corner to catch him over and over again.

Eventually, he was galloping towards the jump. He cleared it. Half an hour later, they managed to manoeuvre him in front of the pole again. Knees snapped up, fetlocks tight to his forearms, forearms close to his neck, neck arched, back flicked up, hind legs kicked up to the sky. They put a few blocks on the barrels, then two more barrels. He was galloping towards the jump again. It was 7 feet (2,1 metres) high at this stage. (The world record for a horse jumping a single pole is 8,1 feet [2,47 metres]). He cleared it with ease. The boys were sweating, skiing around the field at the end of the lunge rope.

They put two more barrels with a single pole in front of the jump, about 4 feet (1,2 metres) away, with another single pole on top. It took another while to get him going in the right direction. When he caught sight of the fence, he crouched low like a cat ready to pounce and cleared the massive obstacle with ease.

The final jump was a seven-foot-high, seven-foot-wide spread.

I leaned against the stone wall, watching and listening to the men swearing and cursing. Then I heard a clear voice, 'Buy this horse'.

I believe that message, wherever it came from, saved me from the ultimate fate of depression. At the time, though, the thought seemed irrational and impossible. We were living hand to mouth and sometimes there was no money for food. The farmer wanted at least fifteen hundred Irish pounds.

When I had arrived on holiday (a year before) I had brought traveller's cheques, I had a part-time job and had saved money for a trip away. I interrupted my travelling to work for a horse dealer. It seemed like a dream opportunity to a pony-mad kid. He never had enough money to pay me, but I didn't mind as I just really wanted to ride all the horses. My bank account quickly emptied when I realised he couldn't afford horse feed either.

I took the bus into Galway City with the baby. At the time there were Foreign Exchange booths around the central square for tourists. I went from booth to booth cashing cheques until I had gathered the price of the horse.

Later that week, somehow, my father managed to contact me to say that the bank had rung him and told him someone had stolen my traveller's cheques and had spent them in Galway. I told him it was me and that I would pay it back. The overdraft sat there for a while…

I bought my horse and named him Coutts, after the bank!

It took time to train him. He kept galloping back into the yard or throwing his rider and taking off down the road to hide in some grassy boreen (a narrow country road).

After hard work and perseverance, he started going to shows. He was a brilliant jumper, but getting him pointed toward the fences was a challenge—he was near impossible to hold or turn.

I have an old video of him show-jumping: we jump a fence, disappear off screen for five minutes (he ran away, around the show grounds) and reappear to finish the course.

Sometimes he would take off a full stride before he needed to. It took him a good while to figure out how to manage the strides in doubles and triples when show-jumping.

When Coutts started eventing, it was terrifying to ride a set of bounce rails with him. His jump was so massive, he would land on top of the second fence. We fell three times while he worked all this out: twice at bounces and once on the flat. On the flat, he hit a soft patch of ground at a flat-out gallop and somersaulted; the fall broke my collarbone.

His motto, 'if in doubt, go faster!' stood to him as he climbed the ranks at events and the fences got more imposing.

Someone offered me fifty-thousand Irish pounds for him when he was eight years old. I refused to sell.

By age twelve, he was 6th in the world three-day event horse rankings.

Coutts and I communicated telepathically, and I learned to silence my thoughts so that he could not pick up on my nerves as we galloped towards some terrifying obstacles.

Sometimes he would observe the fence in the distance and I would hear him thinking, 'I don't know, I don't like it.' I would secretly agree, knowing that high rails at angles over a vast ditch could put us both six feet under in a heartbeat. But I brushed off his concern, 'No, it's fine, nothing to worry about!' Between hands, legs and mind, I steered him—focusing on the exact spot over the far side of the fence where we needed to land.

He could do some incredible athletic manoeuvres, putting down a foot to kick off the top of a fence, flicking his front legs up when too close, taking off 15 feet too soon from a large spread…

While being counted down at the start of the cross-country at an advanced event in Tyrella in Northern Ireland, one of my competitors (bankrolled with expensive horsepower and equipment) called across to us, laughing, "Oh no, not the pony again!" Coutts was cleaning up wherever he went.

He remained impossible to hold on a lead rein, which made the trot up at events a challenging affair. At the Punchestown three-day event, he pulled away and circuited the racecourse before ever starting on his first day of dressage. It was to be followed by roads and tracks, steeplechase, cross-country and finally, show jumping. He had to be ridden for at least two hours before he would behave in his dressage test.

At that time, there was a minimum weight of twelve stone (76 kg) for events. I was only seven and a half stone (47,6 kg), so Coutts carried a lead pack weighing four stone (25kg) under his saddle. I could barely lift it up on him.

He was also impossible to catch in the field. A few people were needed to herd him into the stable. If you caught him, he would pull the rope and run away. No one could hold him.

No one could catch him—apart from one time.

After I had left my partner, I borrowed a trailer. My heart was pumping as I staked out the farm and waited until he drove down to the village. As soon as he was gone, I walked out into the field with a head-collar. Coutts stood still as I put the head-collar on him. He walked effortlessly with me out of the field and up into the trailer.

After finishing his eventing career, Coutts gave a lot of pleasure to young people, winning working hunter and show hunter classes, until well into his 20s.

He was 33, a great age for a horse, when it was time to say goodbye. He wasn't absorbing his food anymore, and you could count his ribs through his blanket of orange fur. It was the end of his last summer. The vet and I led him out to his favourite paddock and offered him a bucket of feed. The vet explained that the injection would make him drop slowly to his knees, lie down and drift off to sleep.

But that was not his nature. He reared up in defiance and fell back, dead.

RIP Coutts (1986-2019). He saved my life on many occasions.

NATURE LOVER INTERVIEW: GORDON DARCY

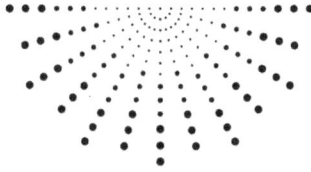

*I would recommend that parents (and teachers) do something
to expose children to nature 'first-hand'.*

— GORDON DARCY IS A LOCAL ENVIRONMENTALIST AND ARTIST.

HOW DID YOU BECOME A NATURE LOVER?

I have been acutely aware of nature since childhood,
particularly birds. I was known as 'Birdy D'Arcy' in school.

I spent a lot of my leisure time watching birds and drawing
them, collecting feathers and wings and looking for birds'
nests.

WHAT IS YOUR FAVOURITE PLANT OR ANIMAL?

I try not to favour one aspect of nature over others, though I
naturally admire birds—particularly birds of prey.

WHAT CAN PEOPLE DO TO HELP NATURE?

Apart from planting a tree, which is always a good idea, I
would recommend that parents and teachers in primary

schools do something to expose children to nature 'first-hand', rather than trough technology.

IF YOU HAD A MAGIC WAND, WHAT IS THE ONE THING YOU WOULD YOU DO TO HELP THE PLANET?

> I would request that every country on the planet would commit to an ongoing tree-planting programme that, where possible, would involve native trees.

Visit www.naturemagic.ie (or search 'Nature Magic' on all major podcast platforms to listen to the full interview).

BIRDS OF THE BURREN

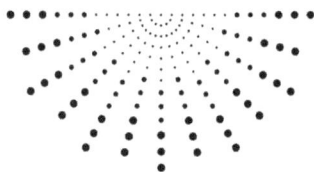

INTRODUCTION

Wild birds are beautiful, but as they are on a constant lookout for predators, they are naturally shy. At the Disappearing Lake (a three-acre, freshwater turlough that fills every 11 to 12 hours!), we have a bird hide. From there, you can see overwintering birds like the colourful, wild teal ducks from Siberia or the majestic whooper swans from Iceland.

It takes patience and a reliable pair of binoculars to spot wild birds. For most people, once they start, bird-watching it becomes an enriching, life-long hobby. It is easy to get 'hooked' on identifying local birds; you will always have something fascinating to focus on.

Getting people interested in wild birds does not have to be hard. Put up a bird table or some bird feeders and buy a bird book. If you get bird feeders, make sure they are away from bushes so that cats can't use them for cover to hunt. Sadly, the domestic cat devastates urban small bird populations. Another fabulous (and easy) activity is listening to and identifying birds' songs and calls. You can even download apps like Song Sleuth and Warblr on your phone. Listening to birds can mend a fractured soul.

EXAMPLE ACTIVITY: WATCHING BLUE TITS REAR THEIR CHICKS

Note: this activity is suitable for nature centres and for individuals who want to do it at home.

The best way to get people interested in wild birds is by using a web-cam bird box. It is a brilliant invention. You get to offer birds (i.e. blue tits) a comfortable home to rear chicks, while at the same time, you and your visitors get to be amazed at the events that follow.

The entire experience is intoxicating! You probably won't see the male when he first checks out the nest or the female when she visits to approve it. But keep watching your bird box every day… The excitement at the first bits of moss in the box is just the first surprise! You will marvel as the female brings in material to construct a perfect, tiny, soft nest. She will weave bits of wool and hair into the inside of the little bed so that it is comfortable for the new arrivals.

Then… A tiny egg! As the days pass, she will keep on laying eggs (her clutch can comprise 6 to 13 eggs). You might see the eggs when she leaves for brief trips (if she doesn't cover them when she goes). Her partner will defend the area around the nest, chasing away anything that could threaten their food source. He will also diligently bring her tasty worms to eat. Two weeks later, perfectly timed with the season when food is most abundant, you will see tiny blue tit hatchlings.

Then both parents start their relentless job of feeding the demanding chicks. They will mostly bring caterpillars, and on the odd occasion a giant slug. It will amaze you to see how the chicks turn up their tails and give their waste sacks to their parents to remove and discard! This keeps the nest hygienic and free from disease.

The hatchlings are pink, ugly and bald. Their beaks seem oversized and as if they have extra hinges. When they open their beaks to beg for food, they resemble the gaping mouth of some alien creature. You will hear them calling whenever they think their parents are nearby.

You will soon see feathers starting to break through the skin. This is obviously an uncomfortable experience as the chicks scratch and peck at themselves continuously. Within a few days, they will turn into cute, fluffy little balls. Soon after, their wing feathers will start growing, and the chicks will begin stretching and fluttering around the box, banging into the ceiling while practising their skills!

The fattest chick will be the first to fledge. He will generally go all the way to the edge of the entrance hole to be first in line for food. Eventually, he will become curious and peek outside. Then he'll fly. His brothers and sisters will start leaving one by one. The parents will return with worms until the weakest chick is ready to fly away. You may, if you're lucky, see them hopping out and landing on a nearby bush. Even after they have fledged, their parents will continue to feed them until they mature, so they are nice and plump for the following winter.

Mother Nature has it all worked out!

LESSONS

- You can show the whole drama on a screen in your kitchen, in a café, classroom or play area. If you have a website, you can stream it for people to watch online.
- It is addictive to watch. You will be compelled to check progress daily!
- The intelligence of wild animals and their ancient evolution that led to these skills being imprinted from generation to generation amazes everyone.
- We experience the miracle of a tiny egg becoming a perfect, fluffy bird. We realise that we have no control over this; this magic is happening all over the world without us even noticing.
- We hold a new respect for the tiny bird who must face this immense task every year. Blue tits will only lay one clutch a year (unless they lose a clutch). The great tit often rears two broods; the female great tit incubates her eggs without takeaway dinners being delivered from her partner (although he may help when the chicks are born). We had a great tit that nested in our web-cam box this year. She reared her brood solo. A stellar achievement!
- Get your box up in late winter. Ours has a wire that connects to a TV inside. Setting this up was an awkward operation, but I believe there are wireless versions. Once set up, it is problem-free.
- The recommended direction to face a bird box is between north and east, as this will provide natural protection from direct sunlight, wind and rain. It will create a more suitable and safer

environment for growing birds. The box can also be tilted slightly forward to allow rain to run clear of the entrance.

EQUIPMENT

Buy your webcam bird box. You can order one online. Ours came from Lidl.

COMMON BIRDS IN THE BURREN

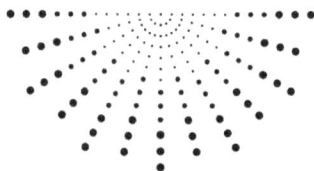

THE ROBIN

Robins used to follow wild boar and ate worms as they turned the ground. Nowadays, there aren't so many wild boar around, but there is still something turning the earth: the gardener. That is why there are so many classic shots of robins sitting on shovels!

The robins at the Burren Nature Sanctuary are used to visitors and sit on branches close to the path, watching people go by. We usually tell the small children to look for Santa's robin. He takes messages back to Santa to tell him how good they have been. They are pretty much guaranteed to spot one around the Fairy Woodland.

In springtime, the fledglings gather in 'gangs' before they disperse to their own territories and defend them ferociously.

THE PEREGRINE FALCON

The Peregrine lives up on the high Burren Hills. We can sometimes see the Peregrine passing high over the farm. He has the best vision on earth; he can see to the farthest horizon and focus on the smallest mouse from over 3280 feet (1 kilometre) high in the sky.

The Peregrine Falcon symbolises the depth and clarity of vision that we should aspire to. We need to focus clearly on the individual decisions that we take, but also to look at the big picture to see how our choices affect the environment.

THE CUCKOO

The Burren is a stronghold for the clever Cuckoo. She arrives from Africa in early spring and lays her eggs in other birds' nests. She enjoys her summer in Ireland while other little birds rear her chicks!

The cuckoo comes in April, sings in May, in June it changes its tune and in July it flies away.

THE WILD SWANS AT COOLE, A POEM BY W.B. YEATS (1865 - 1939)

The trees are in their autumn beauty,
The woodland paths are dry,
Under the October twilight the water
Mirrors a still sky;
Upon the brimming water among the stones
Are nine-and-fifty swans.

The nineteenth Autumn has come upon me
Since I first made my count;
I saw, before I had well finished,
All suddenly mount
And scatter wheeling in great broken rings
Upon their clamorous wings.

I have looked upon those brilliant creatures,
And now my heart is sore.
All's changed since I, hearing at twilight,
The first time on this shore,
The bell-beat of their wings above my head,
Trod with a lighter tread.

Unwearied still, lover by lover,
They paddle in the cold
Companionable streams or climb the air;
Their hearts have not grown old;
Passion or conquest, wander where they will,
Attend upon them still.

But now they drift on the still water,
Mysterious, beautiful;
Among what rushes will they build,

By what lake's edge or pool
Delight men's eyes when I awake some day
To find they have flown away?

NATURE LOVER INTERVIEW: PÁDRAIC FOGARTY

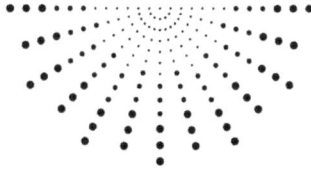

We have to talk about bringing wolves back to Ireland.

— PÁDRAIC FOGARTY IS AN ECOLOGIST AND THE CAMPAIGNS OFFICER WITH THE IRISH WILDLIFE TRUST.

HOW DID YOU BECOME A NATURE LOVER?

I was lucky enough to have grown up near Phoenix Park in Dublin. And at the age of eight, nine, ten, it really was an enormous wilderness. It has some great wild areas.

I remember at the time, it had a nature trail through the Furry Glen. I remember my mother bringing me down to the nature trail and we picked up the map from the visitor centre (that doesn't exist anymore). They'd have things to find on it like fir cones, mushrooms or squirrels or whatever. I really enjoyed doing that.

I had a teacher at the time as well, I think it was in fourth class, who was absolutely obsessed with nature. She made sure we grew up knowing the names of trees and animals and plants and poems and all that kind of stuff.

WHAT'S YOUR FAVOURITE PLANT OR ANIMAL?

> I've always been fascinated with whales and humpback whales, in particular.

They are beautiful animals. They travel such long distances. And I've always been fond of travelling. I've always imagined that if I had to be another animal, I'd be a humpback whale. I would go to the Arctic in the summer.

DO YOU FEEL SPIRITUALLY CONNECTED TO NATURE?

> I don't have to be off in the wilderness or staring out to the ocean to feel connected to nature; I sometimes enjoy watching the spiders and the little blue tits in the back garden, or the way things grow.

I have an elder tree in my garden that planted itself. And it's an entire ecosystem all by itself. So I do feel that connection.

WHAT CAN PEOPLE DO TO HELP NATURE?

> One of the most powerful things you can do is to talk to people around you in an open and respectful way.

One of the reasons I wrote my book was because I didn't feel people really understood the extent to which nature has disappeared from our island. If you don't have a picture of the full extent of the bad news, you're not going to be motivated to act proportionately to it.

I would say the number one action on my list would be to create a nature conservation agency in Ireland that was able to do its job, that was independent, that was able to present the science, to communicate the science about what's happening and to be able to come up with advice for farmers or fishermen, for people, for communities, for people living in towns and cities.

IF YOU HAD A MAGIC WAND, WHAT IS THE ONE THING YOU WOULD YOU DO TO HELP THE PLANET?

I would transform people's relationship with nature. If we don't change our relationship with nature, we're not going to fix the problem we're in at the moment.

What I mean by that is, addressing the feeling that we all seem to have, that we are more important as a species than all the other creatures around us.

We've grown up with this sense of dominion over the natural world. Like, we are in charge and we allow other species to exist. And of course, frequently decide we can't allow other species to exist. Wolves, for instance. For centuries, wolves have been persecuted and vilified.

I know this is very emotional for people. You know, we've grown up believing that farming is absolutely essential. We absolutely need farming, but not everywhere. We need to have working ecosystems with all the species in some kind of balance with each other.

Visit www.naturemagic.ie (or search 'Nature Magic' on all major podcast platforms to listen to the full interview).

Buy Pádraic's book, Whittled Away: Ireland's Vanishing Nature, here: https://amzn.to/3g3DIZV

WILD ANIMALS OF THE BURREN

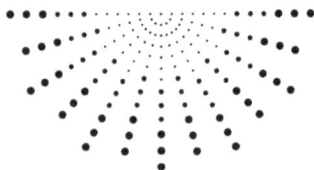

At the Burren Nature Sanctuary, we have 25 acres (10 hectares) of undisturbed 're-wilded' wilderness. This area of shattered limestone pavement, bushes and trees is a haven for Burren Fauna.

THE FOX

Our most abundant wild animal population is the fox. But the poor, misunderstood fox is not popular... If the opportunity arises, she won't hesitate to take a weak lamb home for dinner. And she is always checking out the henhouse to see if, just maybe, you forgot to put the chickens in. But the truth is, she mostly survives on berries, rodents and snails. Foxes are essential in the ecosystem—they keep the rodent population in check.

Sadly, local farmers leave out poison for foxes, mice and rats. When rodents die from the poison, foxes and owls eat them; and they also die. But when you kill off the top predators, the entire ecosystem suffers. The best way to control mice (if you have a problem in your barn) is to have a farm cat. He will happily hunt down the rodents; no foxes or owls need to suffer deadly consequences.

Our lovely member of staff, Emer, rescued a female fox and named her Petri. Petri lived with Emer for a few years. She would come in and pounce on her bed at 3 in the morning! She loved strolling along the lanes for a walk with the family dog. Gradually, she started spending nights away from home and eventually went back to the wild.

If you get to know a fox, you soon realise they are just orange dogs with fluffy tails.

Humans feel entitled to pursue them for sport, hunt them, dig them out and poison them. They have become more tolerated in towns like London, where they trot along the pavements at night with high confidence.

My mother's elderly neighbour in Hammersmith found a fox on her bed, chewing her handbag; it had hopped in through the window. Foxes love eating odd things, especially leather. If you leave a head-collar out in the field, you will find the buckle chewed away the next day.

One of our resident foxes is obsessed with shoes. We often find single wellies, Crocs and runners half-chewed in the field. We can only imagine he has robbed them from our neighbours' porches.

THE PINE MARTIN

The elusive pine marten, or cat crainn in Irish ("tree cat"), is the nightmare of poultry keepers. At the sanctuary, we have seen her skulking along the tops of stone walls. A visitor has photographed her kits (babies) while they were playing around the nature walk.

She is rarely spotted, but recently she was seen walking high up along the canopy of some tall sycamore trees, checking crows' nests for eggs. We have set up a wildlife camera pointed at a peanut butter sandwich (apparently their favourite!) to see if we can get a photo of her.

It is difficult to keep the pine marten out of your poultry pen. Here is a link to a useful leaflet that shows you how.

https://www.vincentwildlife.ie/wp-content/uploads/2015/04/Pine-Marten-Leaflet.pdf

THE BADGER

The 'rocks' at Burren Nature Sanctuary (15 acres [6 hectares] of shattered limestone pavement) are home to another persecuted animal, the badger. People blame badgers for carrying TB around; they are often dug up from their dens by hunters or poisoned by farmers. There is now a vaccine program for badgers. Hopefully, this slaughter will stop.

Badgers are beautiful animals. They are nocturnal and have a strong snout like a pig with which they dig setts (underground dens) and root around for snails and worms. They have powerful jaws to bite through sticks and roots. They are excellent diggers, and live underground in setts that often get passed down from generation to generation.

Once, we got a call from a local vet saying that someone had handed in a badger. She had been run over and needed a home to recover. She had a few scratches and also a head injury. The vet thought she wouldn't survive back in the wild. Typically, badgers are very vicious…

"She is very quiet, she's just wandering around the kitchen and sipping milk. You can stroke her and hold her. Can you take her?"

It took a few days to look into getting a license to give her a home as she was a wild animal.

When I rang the vet back, she said, "Oh, we released her. She came to her senses, and we let her go. She was very cross. She must have been concussed!"

THE IRISH HARE

The hare is a magical creature of myth and legend. In Irish mythology, they believed the Irish hare represented Eostre, the goddess of the Moon.

The native mountain or Irish hare, (*Lepus timidus hibernicus*) is one of Ireland's oldest mammals. In County Waterford, they have dated Irish hare bones over 28,000 years old.

It is an honour to see a hare galloping across a field or clearing a stone wall in one jump.

Years ago, we found a leveret (baby hare) under the tractor in the yard. We thought it was lost. It was the most adorable creature ever. We brought it in and kept it in a rabbit cage overnight. After some research, we learned that the mother hare places her babies in different spots around the area so they are not all vulnerable together. She then goes around to feed and check on them in rotation. Under the tractor wasn't an ideal spot. But we put him back early the next morning to wait for her and said our sad goodbyes to our sweet little friend.

The Irish hare also graces the Irish pre-decimal threepence piece.

HEDGEHOGS

One spring, a lady arrived with two hedgehogs. They were weak and very light. She found them in a round bale of hay when she started unwinding the layers to feed her sheep. We kept Hotchy and Witchy in the stable for a few weeks and fed them until they were healthy. They were charming. It was a sad day when we released them back into the wild to take their chances. A month after their release, I met one crossing the field, so I hope they fared well.

Hedgehogs have large territories and roam up to a mile (1,6 m) every night. They are suffering because their habitats are being 'eaten up' by little gardens and roads.

This year, we undertook a hedgehog survey at Burren Nature Sanctuary. Sadly, we did not record any hedgehogs in the area, but it was interesting to see the method. You staple two pieces of paper to the bottom of triangular tunnels, and paint oil mixed with charcoal in the front. If an animal walks on the black 'paint', you will see its footprints on the paper. We recorded our nosy cat, some rodents and slugs! Apparently, if you have a lot of badgers in the area, it is unlikely that you will see hedgehogs.

How you can help hedgehogs.

Make a hedgehog door into your neighbour's garden and leave a shallow dish of water out in hot weather. They eat dog food if you want to leave some out. Never give them milk! They love it and will drink it, but the milk doesn't like them at all.

FERAL GOATS

The Burren is home to hundreds of wild goats. They are not popular with farmers as they can eat a field of grass overnight. They are descended from escaped domestic goats that have been farmed in the Burren for over 4,000 years.

Our rescued mountain goat, Frisky, is a Burren feral goat. She has long horns and a long, shaggy brown and white coat.

BATS

Every Halloween, we look for an appropriate 'spooky' animal theme. We have chosen owls, spiders, worms and, of course, bats. For the bat theme, we approached the local bat society. This inspired us to get a bat detector.

Echo-location was only discovered in the 1950s. Bats emit ultrasounds (too high for humans to hear) that bounce off objects, which allow them to fly in the dark.

You can use a bat detector to convert their calls into audible sound. Different species emit sounds at different frequencies. A cheap bat detector can tune into different frequencies. You can use the device's handbook to discover which species' sounds it picks up when using it.

We tested our bat detector the first evening we received it. I opened the back door and stepped out in my pyjamas. Roy, my husband, watched from the doorway. I turned on the bat detector and instantly tuned in to the squeaking. Bats were flying all around me; Roy said I was some kind of bat Dr Doolittle. We identified pipistrelle, a common micro-bat in Europe and one of Ireland's smallest mammals.

In Ireland, it is an offence to disturb, injure or kill bats. The law protects them. Visit this website for more information: https://www.batconservationireland.org

The bat quiz is especially fun, as bats have so many amazing, fun facts about them. See 'Nature Quizzes' in the toolkits section (page 207).

Another one of our podcast guests, Sean McCormack, advertised a bat walk in London and over 400 people applied! This led to him setting up the successful Ealing Wildlife Group. (https://ealingwildlifegroup.com)

Other Burren Fauna include lizards, snails, frogs, red squirrels, stoats, weasels, rabbits and small rodents.

Seeing a stoat is a good omen. In the words of Patrick McCormack, "Even the little stoat that crosses the road is a sign that luck is coming my way financially!"

There are no snakes in Ireland. But in the Burren, we have the slow worm, a legless lizard that could be mistaken for a snake. Apparently, it was introduced in the 1970s when it escaped from a hippy!

LESSONS

- Wild animals are, not surprisingly, elusive. They mostly move around at night. But... we often see our resident foxes sunbathing in the day. They seem to know they are in safe quarters.
- It is easy for people to justify putting down poison for a fox that came in the night to steal their chickens. But if they could witness the same fox lovingly trying to rear a litter of five or six cubs and killing up to six rats (another important but unpopular species) a day, they may become more understanding of the bigger picture, and more tolerant towards this species.
- So, please spend money on securing your henhouse instead of buying poison to wipe out the fox population and with it, owls, hawks and peregrines.
- Each animal has a vital role to play in the ecosystem.

THE WILD OTTER RESCUE AND RELEASE PROJECT

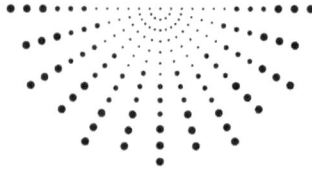

We have seen otters at the Disappearing Lake at our sanctuary. They live downstream, about 0,62 miles (1 km) away, where the large underground river appears nearby Dunguaire Castle in Kinvara Bay.

On my 50th birthday, my husband borrowed a kayak. Everyone said there was no chance of seeing one, and they probably weren't living there at all… We paddled from the Quay around the Castle. As we were paddling, a little head popped up in front of us and swam around. It was a young otter. He got up on the bank, had a good look at us, and slipped into the undergrowth—pure magic!

One or two baby otters get rescued in Ireland every year. It is notoriously difficult to rear an otter correctly. They need looking after for a full year because they cannot survive solo until they are fully mature. Also, they must not become too friendly with humans as that could lead them to danger upon their release.

Inspired by two amazing retiring sanctuary owners, Clive and Judy Laurence in Connemara, we hope to construct an Otter Rescue and Release Facility. If built near the turlough (our Disappearing Lake), the otters would become familiar with their territory and hopefully find a new home up- or downstream when they are ready for release.

Otters need a place to swim. Clive and Judy have a tank, approximately 21 square feet (2 square metres), for the baby otters to practise swimming in. They also need a clean, dry bed with a way of isolating them when you clean it. They eat at least one kilo of fish per day.

When we visited Clive and Judy, they let their young otters out to play in the garden. We sat in the kitchen and watched as they raced around and knocked over plant pots. Because Clive and Judy live close to the sea in a remote area, this is ideal. Eventually, the otters get familiar with their surroundings, and they are released into the wild. As there are no houses between them and the sea, the released otters can return for food (if they wish) until they have worked out how to survive on their own.

We would like to build a wild otter rescue and release facility to provide this service.

MAGICAL STORY
WILD HERB OF THE FLAT ROCK

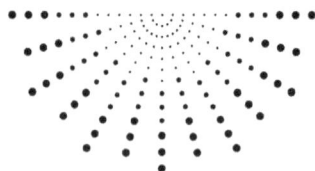

This is a story based on real-life events.

Three light knocks. Tap, tap, tap.

Jane sees a leathered face peering around the door. Deep wadis run down tanned cheeks—gullies worn away by the tears of the ancestors. He takes off his dark, wide-brimmed felt hat. His shiny blue-grey hair is tied back. Thick ropes of turquoise beads hang over his neatly pressed charcoal shirt.

"Yá'át'ééh abiní. It is good, the morning." He says.

She lifts her head. *A dream, a trip, reality?*

"In beauty, I walk." He puts his hands together in prayer.

She didn't change clothes to sleep in. She wore layers of shirts and skirts. Her kids wore t-shirts or jumpers, and bare arses or jeans—depending on their potty-training level. She wrapped them in cloth nappies at night.

She pulls on heavy boots, her brown, frizzy hair is matted in clumps. The house is warm. It smells of patchouli, Cannabis and wood smoke. It is early; the babies stir under their blanket in the corner like a litter of pups.

"My name is White Deer of Autumn," the man says softly, "I am a Shawnee elder. We were passing through and were drawn to your round house."

Two other elderly men with long plaits wait behind him. Jane stands frozen for a minute, the image searing into her core memory. The early morning sun paints the stone walls pink and gives the grass a neon green hue.

The birds sing raucously. Jane observes the fire pit to the right, the poly-tunnels, and the painfully empty caravan beyond them. Her attention turns back to the three chiefs.

It is a sign.

"We wish to offer thanks to the people of this land," White Deer of Autumn gestures to his elderly friends with an open hand.

A smaller man with a broad, flat face; wearing a Polo Sport windbreaker over a denim shirt and jeans with a large, silver belt buckle steps forward. A feather is woven into his shiny, black braid.

"Chief Little Wolf Stares at the Moon, from the Cherokee people," says White Deer of Autumn.

"May the Great Spirit walk with you." The second man says, gesturing towards the ground.

An ancient-looking man with a tasselled coat, glassy blue eyes and an attractive, dangling earring leans on a high staff.

"Chief Soaring Singing Hawk from the Mohawk people." Says White Deer of Autumn.

"May the warm winds of Heaven blow on your house." He kisses his palm and blows on his hand towards the thatched roof.

"Come in. I'll put the kettle on."

She pulls the crunchy, square towelling nappies off the line and throws them in a basket.

The three men arrange themselves on cushions around the ashes of the central fire. She lights some paper and a few small sticks and then adds some logs.

Peace falls on the house like a thick, warm Navajo rug.

The fire blazes up in front of them, heating their cheeks while the cool breath of morning refreshes the back of their necks. A pair of wrens serenade them from a Buddleia bush. The sweet scent swirls in tantalisingly through the open door.

A pine marten trots along the top of a wall. A wave of feral goats passes across a rocky field in the distance, their musk briefly tainting the air. A donkey brays across the hills, welcoming the day.

"We landed in Shannon yesterday," explains White Deer of Autumn.

"An American bald eagle got caught in the Gulf Stream (a warm, swift Atlantic Ocean air current) and then flew over the Atlantic and ended up here, in the Burren. Your wildlife wardens caught him and put him on a flight from Shannon back to New York. They released him somewhere along the Mohawk River. We are thankful for your kindness. It is a sign that this place is a special place for nature. We come to give thanks."

Jane fills the kettle with water from a five-gallon plastic churn.

Drawing water from the well and wheeling it back to the house is a tough job in summer, and torturous in winter.

Brad had been good at that. He was good at gathering firewood, too. He was good at keeping her warm. She was forty when she met him, forty-four when he left her with four young ones under the age of five.

He was a blessing... And a curse.

The last twenty years reel through her head: leaving London, joining the new-age travellers, buying the patch of land with Mikey, building the round house stone by stone, planting the garden; year after year trying for a baby, him leaving, his blond, buxom new woman and their clutch of kids, her romance with Brad, his black and white horses, her and Brad's babies born next to the fire, him drinking, fighting, leaving, and finally, her facing the looming winter alone.

She throws handfuls of mint and lemon balm into the teapot and adds some raw honey.

The kids had been up playing in the evening until the sun had set at 10 PM. They stir a bit but settle back to sleep, well used to the murmur of adult conversation around the fire.

She cuts slices off a brown soda loaf. It is a little stale. She toasts it and spreads her home-made blackberry jam over it.

"Thank you for the food." Soaring Singing Hawk bows his head in gratitude.

"We would like to perform a blessing ceremony for you, but first, could I ask you to introduce us to your native flora and fauna? We are not familiar with your red cattle, your small robins and little windblown trees. We would love to learn their names and tell our people about them when we return."

Jane agrees, and they go outside. They walk through the herb garden. She shows them the polytunnel, bursting with produce. Slowly they pick their way across the limestone pavement, observing the sunken beauty of Mullaghmore in the distance; water is shining off the turlough at her foot.

A pair of peregrine falcons circle high above them, on guard and searching for prey.

"The windblown trees are mostly whitethorn. The leaves can help regulate the heart when you chew it. The other little bush is called blackthorn. It has a bitter berry called a sloe which is sometimes used to flavour gin and vodka. Hazel trees give us nuts. The cattle are shorthorn. They graze the mountains in winter, when the springs are full, and go down to the meadow pastures in summer."

Little Wolf Staring at the Moon points at a wild rose, "Spikes instead of thorns?"

"The burnet rose, it makes black hips in autumn."

They stop at a little, lumpy hill. "Ants," says Jane. She pulls a few leaves of thyme off the anthill.

"They garden the herbs to protect the hive against sickness. Thyme is an antiseptic. See how short they trim it. They have wood sage, wood sorrel and oregano."

They step over deep, narrow crevices of rock, past patches of grass and flowers.

"This is the fly orchid, the bee orchid and the frog orchid. These are mountain avens from the Arctic. And this is the maidenhair fern from the tropics. There is no discrimination on the mountain!"

She points out butterflies, moths, a slow worm, a hare, foxes and badger tracks. They stop at a little stream where some enamel mugs are hanging on a bush. They drink the calcium-rich water in silence like the ancient holy men from long ago, finding the Almighty in the empty spaces. Time stands still.

When they return, the children are awake. They're drawing shapes with sticks in the fire pit ashes. Three of them run to Jane, screaming with glee while grabbing her skirts. The baby sits on the threshold of the house, bawling for food.

"Rowan, can you get us the cushions, my love?" She asks the biggest child, who is gnawing on the heel of the bread. She scoops up the baby and sits, breast-feeding her two youngest children in turns as she talks with the three chiefs. The two older children work the raspberry bushes in the hedge, and run to and fro, bringing berries to the old men.

Rowan brings an ember from the house fire and starts blowing on some dry grass. When it catches, he piles up little sticks and gets the outside fire going for his mother.

The elders start to drum a beat on their knees.

Jane puts the baby in a hammock, made from a shawl tied between two bushes, and rocks her occasionally. Rowan runs into the house and brings out a bodhrán and Brad's bongo drums.

He was good on the drums, and the guitar; a lovely singer.

The three toddlers join in with sticks.

White Deer of Autumn unpacks a pinkish stone pipe and slowly fills it with whole leaf tobacco. He sprinkles some tobacco on the ground.

"To acknowledge, we must always give back to Mother Earth."

He takes the feather from Little Wolf's hair.

"A gift to you—a bald eagle feather from our country. It will give you courage and strength. Think of us when you hold it. He will be your guide. If your children are sick, I will give you healing heat. Close your eyes, and you will see me blowing on the embers like that child, and the heat will surge from your hands. Your hands will know where the healing is needed, and the heat will pulse from you. Little Wolf and Singing Hawk will stand guard on your space as you work. To the right and left of the sick, at their shoulders."

He brings the pipe over to the flame and lights it.

"We call upon and thank the six energies!" Holding the bowl in his hand, he points the pipe to the west.

> Black is the colour of the west
>
> Where the sun goes down.
>
> Black is darkness, release, spirit protection.
>
> Black is the cup of water,
>
> The life-giving rains come from the west.

White deer of Autumn turns and points the pipe to the south.

> The south is yellow!
>
> Our Mother Earth gives us growth,
>
> And herbs that heal us.
>
> We think of her strength,
>
> We think of her bounty and physical healing,
>
> While we load this pipe.

He turns to the east.

> Red is the east!
>
> It is where the daybreak star,
>
> The star of knowledge appears.
>
> Red is the rising sun,
>
> Bringing us the new day.

He sprinkles a little tobacco on the ground and turns to the north.

> White is the north!
>
> The North covers our Mother Earth
>
> With the white blanket of cleansing snow.
>
> The winter is the time of long contemplation
>
> For us two-legged.
>
> When we have the face of the old,
>
> We will want to look back upon our lifetime
>
> And hope we stood for the straight road
>
> In our relationship to all things.
>
> Courage and enduring strength, truthfulness and honesty.
>
> These strengths we seek
>
> As we stand here facing north.

He holds the pipe above his head.

> May the sun bring you energy day by day!
>
> You have earned a new name.

I call out to the birds, to the plants, to the rocks and the animals,

To introduce them to you.

Great spirit, mystery, unexplainable source of life.

We thank you for the six powers of the universe.

Please keep safe our friend,

Wild Herb of the Flat Rock.

Footnote

The above is a true story that happened in 1989. The bald eagle did get blown over the ocean to the Burren. The three native American chiefs did visit the Burren and knocked on Jane's door (I changed her name for confidentiality) and performed the peace pipe ceremony. I know this because the eagle story was in the paper, and I was the next person to knock on that same door. From what she told me, I imagined how the visit went. I have since found out that the chiefs were from the Crow people.

I was nineteen years old, with a month-old baby, suffering from severe mastitis and a high fever. The doctor had injected me three or four times with penicillin, but it kept coming back. It was so painful with the last visit; I wanted him to put me to sleep.

I met a lady from Dublin who told me to try a comfrey root poultice. The only people I could think of that might have a comfrey plant were the new-age hippies, so I tracked down the band of wagons and tents. They sent me up the mountain to Jane's round house, as they knew she had a large herb garden.

I saw the empty caravan.

I left with the comfrey root and solace for my own situation.

Jane told me how she and her husband had tried to have children for 20 years… And as soon as they split up, they both had a few!

I remember her nut-brown babies rolling around on the grass. How she raised them alone on that mountain, I don't know. They are in their 30s now.

The last time I saw her was five years after she had given me the comfrey root. She was in the queue at the post office in town with her happy, raggle-taggle gang of children.

The comfrey poultice cured the mastitis within three days.

NATURE LOVER INTERVIEW: PROFESSOR JANE STOUT

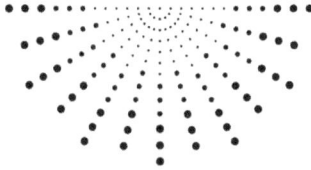

I love bees, but by the end of this, you are going to know that I think all insects are awesome!

— PROFESSOR JANE STOUT IS A BOTANY LECTURER AND CO-FOUNDER OF THE ALL IRELAND POLLINATOR

PLAN AND THE IRISH FORUM ON NATURAL CAPITAL.

HOW DID YOU BECOME A NATURE LOVER?

I was very lucky. I grew up in a little village in the countryside. My bedroom window looked out across fields. And my mum used to teach me the names of wildflowers. My dad would teach me the names of the birds. We would go for walks in the woods. I think it cemented for me when I started studying biology for a level at school and we went on an ecology field course and it just started to blow my mind.

And then I went to university, studied environmental sciences, and it wasn't really until the end of my degree that I really focused on ecology, insects in particular.

This love of nature, it's just sort of grown gradually.

WHAT IS YOUR FAVOURITE PLANT OR ANIMAL?

I am an advocate for insects. There's a million different species, but they're really the under-appreciated animals. You know, they're the most diverse group of animals on the planet. There are more insects than anything else. They're incredibly diverse, successful in every ecological region, every habitat on every continent. My favourite insects are the bees and especially the bumblebees. I did my PhD on the foraging ecology of bumblebees so my favourite would be Bombus pascuorum, the common carder bee.

DO YOU FEEL SPIRITUALLY CONNECTED TO NATURE?

We are all part of nature. There's been research that shows being in nature can give people a sense of peace and wellbeing. There is the science now that shows the level of stress hormones in your body is reduced if you're exposed to nature.

WHAT CAN PEOPLE DO TO HELP NATURE?

I think sometimes we're overwhelmed by some of these environment challenges.

You know, climate change is a global problem. What can I as an individual do? And if I do something, how do I see the results of it? So it's not that we need to save it. It's that we need to leave room for it.

We need to protect the diversity of ecosystems that we have. The first thing is: do less. That's something we've been promoting through the pollinator plan in terms of doing less. I want nature taken into account.

IF YOU HAD A MAGIC WAND, WHAT IS THE ONE THING YOU WOULD YOU DO TO HELP THE PLANET?

> If I could do anything, I think it would be for everyone to appreciate nature and its value. Nature is wonderful.
>
> It's the result of millions of years of research and development, and it's our life support system. And I really just would love to see a stop to the destruction of nature. You need to have nature taken into account in all of our decisions so that it's restored and protected.

Visit www.naturemagic.ie (or search 'Nature Magic' on all major podcast platforms to listen to the full interview).

Check out the All Ireland Pollinator Plan and the Irish Forum on Natural Capital here: https://pollinators.ie; https://www. naturalcapitalireland.com

INSECTS OF THE BURREN

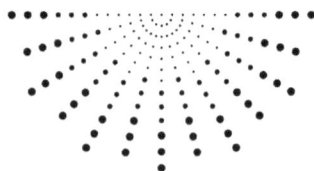

BEES

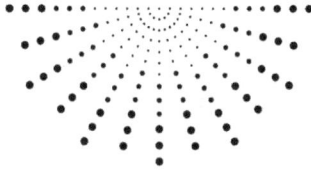

When we opened Burren Nature Sanctuary, we received a grant from Galway Rural Development. In the original application, we applied for a beehive. The hive arrived in a few hundred pieces, and it was two years before I tackled the job of constructing it and making all the frames. Once the hive was complete, I visited a friend's bees, joined the local beekeepers and went on their six-day training course. But we never got around to actually getting bees.

One day, I received a text: "My mother has a swarm of bees on a tree stump in her garden. If you want them, you need to come now!"

Bee colonies 'swarm' at least once a year; this happens when a new queen bee is born within an existing colony. When she flies away to start her own colony, the existing one divides into two, and half the bees fly off with her. Before they find a suitable place to build a new hive, the swarming bees will cluster around their new queen in a ball. You can spot them hanging off branches or in the eaves of houses. While they wait, about 50 'scouts' will fly off in search of the perfect spot to build the hive. Typically, the bees will have found a new home and flown away within a few hours of swarming. Sometimes, an existing hive can experience after-swarms if more than one new queen is born. Such after-swarms can continue until the original colony is all but depleted of worker bees.

I opened my brand-new bee suit, located the smoker and bee brush, and found a suitable box and some tape. We did not discuss how to catch a swarm of bees in the course I had taken; their advice was to contact an experienced beekeeper for the task. I called my bee friend, but he was at work. He gave me a crash course over the phone and off I went.

The bees had lodged in a hole in a low tree stump. We lit a few leaves in the smoker. And after many attempts, I went over to puff smoke over them. Smoke interferes with bees' ability to release alarm pheromones, which alert the colony to danger. By preventing 'guard bees' from releasing these pheromones (and subsequently the same response in the rest of the colony), the bees become much calmer and are less likely to attack. I took the bee brush and attempted to sweep the bees into the box. A quarter of them went in. Mrs Kelly ran off to get another box; I made another attempt and tried to sweep some of them off the ground. Several bees stayed lodged in the trunk, and another lot fell to the ground. I got as many as I could in the boxes and taped them shut.

My bee advisor told me to get a white sheet and drape it from the hive entrance to the ground and pour the bees out on the sheet. I hung the sheet, but as soon as I started opening the taped boxes, the bees began to escape. I opened the box and moved away. They were all flying about, so I left them in the hope they would smell the frame of honey that I had left in the hive to tempt them. I went back to check on them later that day. When I peeked in the box, they had all gone back inside and formed a clump. I opened the lid of the hive and swept them inside. I put a feeder containing sugar and water (which my bee advisor friend had kindly given me) in the hive and left them, fingers crossed.

When I went back to check three days later, they were still there! I must have captured the queen, or they all would have gone searching for her. They are the native black Irish honey bee.

The Solitary Bee

Most bee species in the world are solitary as opposed to social. This means they exist as 'lone' males and females. Once they have mated and prepared a nest for their offspring, they die. Their young will fend for themselves once hatched.

Solitary bees are responsible for a vast amount of pollination. You can try to identify solitary and bumblebees and, if you'd like, give them somewhere to live.

EXAMPLE ACTIVITY: MAKE A BAMBOO BEE HOTEL

Owning bees is an enormous commitment. You must check up on them weekly. And many things can go wrong!

Join your local beekeeping community (and do a course to educate yourself about adequately caring for bees).

Offering a house to a solitary bee is easy!

What You Need:

- Drainage (rainwater) pipe, 8 inches (20 cm) long.
- Bamboo canes, 8 inches (20 cm) long and approximately 0,39 inches (10 mm) in diameter.
- String.

Cut two holes in the middle of the drainage pipe on one side, and insert the string through the holes, so you can tie it to a branch (it will hang horizontally). Next, tightly pack as many bamboo canes as possible into the drainage pipe. And You're done! Tie your newly constructed 'bee hotel' to a tree.

You can also make a bee pallet hotel as explained in The Bee Book by DK, Emma Tennant and Fergus Chadwick. Also, consider planting bee-friendly flowers for all seasons. Visit the Toolkits section (page 206) of this book to learn about running a bee dress up activity for children.

Equipment Needed for Bee Keeping

- Beehive.
- Frames.
- Bee suit.
- Smoker.
- Bee Brush.
- Hive Tool.

LESSONS

- Bees teach us about community. They all have specific jobs for the survival and health of the hive: the queen bee lays the eggs, the drone bees mate with the queen, the nurse bees take care of and feed the brood, guard bees protect the hive by chasing off enemies and fanning the hive if it gets too hot, scouts look for food or places to build and worker bees do the work!
- Bees will only sting if threatened, but they willingly sacrifice themselves for their community.
- If you own bees, you will instantly become a member of the local beekeeping community. They will advise you on everything from equipment to how to harvest your honey. Beware though, beekeepers are obsessed!
- Bees pollinate 70% of global food sources that 7 billion people rely on. That means 4,9 billion people are at risk of starvation should we ever lose the bee species (as a whole).
- We are losing bees at an alarming rate. Although new diseases have been identified that are killing bees, the main culprit responsible for unprecedented bee deaths is the use of herbicides and pesticides.

BEE FACTS

- Bees dance to communicate the source of good flower patches to the colony.
- Bees carry pollen in baskets on their knees.
- Pollination services are valued at an estimated at $170 billion per year.
- Bees are under threat. Colony collapse has been seen in 50% of hives worldwide, probably due to pesticides.
- Bees can see ultraviolet light.
- Bees are sensitive to the electric fields of flowers. As a bee flies, it bumps into charged particles. This leads to friction which strips the bee of electrons, leaving it (the bee) positively charged. On

the other hand, flowers can sometimes be negatively charged. Consequently, when the bee approaches the flower, pollen literally jumps from the flower to the bee.

LAKE ISLE OF INNISFREE, A POEM BY W.B. YEATS (1865 - 1939)

I will arise and go now, and go to Innisfree,
And a small cabin build there, of clay and wattles made:
Nine bean-rows will I have there, a hive for the honey-bee;
And live alone in the bee-loud glade.

And I shall have some peace there, for peace comes dropping slow,
Dropping from the veils of the morning to where the cricket sings;
There midnight's all a glimmer, and noon a purple glow,
And evening full of the linnet's wings.

I will arise and go now, for always night and day
I hear lake water lapping with low sounds by the shore;
While I stand on the roadway, or on the pavements grey,
I hear it in the deep heart's core.

BUTTERFLIES

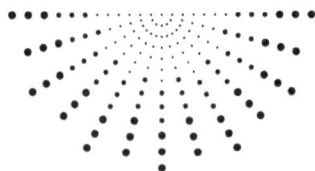

The Burren is home to all but two of Ireland's 30 species of butterfly. They are the perfect 'starter' insect for both children and adults.

A botanist from the National Biodiversity Database came to do a talk about the flora in our meadow. He was instantly jumping around like a ballerina with his butterfly net and identified species like the green hair-streak (*Callophrys rubi*), the small blue (*Cupido minimus*) and even the rare pearl-bordered fritillary (*Boloria euphrosyne*).

There are also over 1,400 moth species in Ireland. Two stunning varieties are the Burren green moth (*Calamia tridens*), which is a treat to see with its stunning lime green colouring; and the gorgeous red-and-black cinnabar moth (*Tyria jacobaeae*) that feeds on the most despised plant, ragwort.

EXAMPLE ACTIVITY: IDENTIFY BUTTERFLIES

Buy a butterfly net and a butterfly identification book. Head off to a nature reservation and see what you find!

Fun Fact: Butterflies taste food by standing on it. This is because their taste sensors are found in their feet.

Moths are just as beautiful to study. Buy an observation moth trap with a light and assemble it on a quiet, dry night. The numbers, colours and variety of moths will amaze you.

The table on the following page displays a list of butterflies in Ireland.

Common Name	Irish Name	Scientific Name
Brimstone	Buióg ruibheach	Gonepteryx rhamni
Brown Hairstreak	Stiallach donn	Thecla betulae
Clouded Yellow	Buióg chróch	Colias croceus
Common Blue	Gormán coiteann	Polymmatus icarus, Polymmatus icarus mariscolore
Dark Green Fritillary	Fritileán dúghlas	Mesoacidalia aglaia formerly Argynnis aglaja
Dingy Skipper	Donnán	Erynnis tages
Grayling	Glasán	Hipparchia semele
Green-veined White	Bánóg uaine	Pieris napi
Holly Blue	Gormán cuilinn	Celastrina argiolus
Marsh Fritillary	Fritileán réisc	Euphydryas aurinia
Meadow Brown	Donnóg fhéir	Maniola jurtina iernes
Orange-tip	Barr buí	Anthocharis cardamines
Painted Lady	Áilleán	Vanessa cardui
Peacock	Péacóg	Inachis io
Pearl-bordered Fritillary	Fritileán péarlach	Boloria euphrosyne
Purple Hairstreak	Stiallach corcra	Quercusia quercus
Red Admiral	Aimiréal dearg	Vanessa atalanta
Ringlet	Fáinneog	Aphantopus hyperantus
Silver-washed Fritillary	Fritileán geal	Argynnis paphia
Small Blue	Gormán beag	Cupido minimus
Small Copper	Copróg bheag	Lycaena phlaeas
Small Heath	Fraochán beag	Coenonympha pamphilus
Small Tortoiseshell	Ruán bheag	Aglais urticae
Small White	Bánóg bheag	Pieris rapae
Speckled wood	Breacfhéileacán coille	Pararge aegeria

DRAGONFLIES

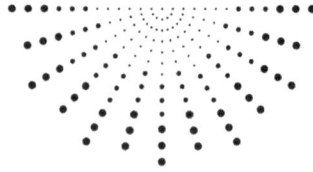

Dragonflies are magical, beautiful creatures. Seeing the neon-blue of a dragonfly hovering over water is as special as spotting your first bright cobalt-blue spring gentian in April on a grey Burren hill.

Dragonflies control each of their wings independently; they are superb fliers and one of the earliest creatures to have developed the ability to fly (30 million years ago).

They eat hundreds of mosquitos and small flies a day, hovering and catching their prey from above with their legs.

Their massive, globular eyes have souped-up colour vision that is better than anything so far discovered in the animal world.

Humans have what is known as tri-chromatic vision, which means we see colours in combinations of red, blue and green. This is thanks to three different types of light-sensitive proteins in our eyes, called opsins. A study of 12 dragonfly species has found that each one has no less than 11 opsins. Some even have a whopping 30 opsins. That means they can see more colours than us!

ANTS

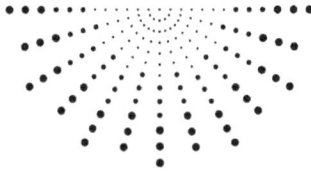

We have lots of anthills in our wilderness area at Burren Nature Sanctuary. I was not especially interested in them until I walked around the Burren Walk with a myrmecologist (ant specialist). The Sanctuary often draws highly qualified nature specialists. It is incredible what you can learn from them.

He stopped at an anthill. It was a small, grassy hillock about 3 feet (0,91 m) high; he told the group that from the size of the anthill, it could be 500 years old. We peered closer. He pointed out what was growing on the hive: wild thyme, wood sage and wild oregano. He explained that ants harvest seeds from the plants they want to use and carry them to their hive.

They remind us that we may not understand the dynamics of different societies, but we must not dismiss them as unsophisticated.

Ants are not as popular as bees, but their communities are just as complex. For example, ants 'farm' aphids. They keep them in herds and 'milk' them for the sweet honeydew they produce. Do not underestimate the intelligence of other species; it may be different, but that does not mean it is less than our own.

Dr John Breen of the University of Limerick roamed the Burren with a knapsack full of breadcrumbs. He scattered them carefully at five different locations, each measuring 32 square feet (10 square metres), and then looked at what turned up (or didn't) in each of the 500 one-metre squares (or 'quadrats') within. At least one species appeared in all but 69 of the squares. He counted 13 different ant species, which makes the Burren, yet again, a national oasis of biodiversity.

He commented, "Today's intensive silage fields, ploughed and reseeded with a ryegrass monoculture and compacted by farm machinery, are no landscape for anthills. Only in places such as the Burren is there the ecological stability that encourages long-lived ant colonies, their dry mounds of fine soil covered in summer with purple flowering thyme."

MAGICAL STORY
ANIMAL TOTEM HEALING

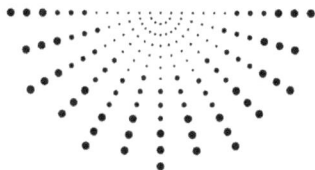

The first year Burren Nature Sanctuary opened was intense. It was a steep learning curve for all of us, from dealing with customers to managing staff, running a café, keeping the entire place clean and maintained, offering nature activities, school tours, etc. It was like having a new, demanding baby. We also had the farm and two children to look after. We were exhausted.

My good friend, Jan, called me. She sensed I was stressed. Jan is a well-renowned therapist and had been using journeying techniques for years. She even treated groups of young offenders in detention institutions successfully.

"Steve is coming over, he's doing a workshop, and you have to come. It will be good for you."

Steve Gallegos was 83 years old and had trained Jan years ago. He was famous all over the world for his healing work and is the author of The Personal Totem Pole Process and Little Ed and Golden Bear.

"He never comes to Ireland. He definitely never will again, and he's doing a workshop in the Lady Gregory. (a hotel in Gort, ten minutes away from the sanctuary) It's €100 for the morning."

I really didn't have time, I was tied to the business and €100 was a lot of money, but…

"You have to come, it's the chance of a lifetime." She persuaded me.

We arrived in the hotel function room. There was a group of about fifteen people. Steve introduced everyone and said that the session was themed on bringing people 'home' to their true selves. He explained how everyone would lie on the floor and how he would start with a relaxation technique, before leading us into the journey. He started speaking in a lovely, low, melodic tone.

"Relax your feet, your ankles, your calves…"

That was the last I could remember. I fell asleep on my back on the thick, flowery carpet. By the time Steve woke me up, two hours had passed.

I was raging inside. I had missed everything. I felt like a stroppy toddler who had missed the ice-cream van! I had spent €100 to sleep on the floor of the Lady Gregory Hotel…

He started moving around the room, asking people what their experience had been like. Each experience was more beautiful than the previous one, I was green with jealousy.

"I was in a forest," "it was a treehouse," "there were animals," "I was lying in a hammock," and so it went on.

He got to a guy in his twenties, a farmer type, who had been dragged along by his girlfriend. He had looked embarrassed at the start, making sure everyone knew that going to the session wasn't his idea. When Steve asked him what he had experienced, he was visibly annoyed.

"Nothing. I saw nothing,"

"Okay, shut your eyes. What can you see?"

"Black, all I see is black. There is nothing, I see nothing. It doesn't work."

"Move back a little," Steve said, "perhaps you are too close."

A few minutes of silence followed.

"Oh, I can see a black horse." His girlfriend's eyes widened.

"A beautiful, shiny, black horse."

"Can you rub its neck?" Steve wanted to know.

"Yes…" A tear trickled down the young man's cheek, "It's my father."

By the time he got to me, I was sulking. I had witnessed none of this magic.

"I fell asleep; I didn't see anything." I said petulantly.

"It's okay, I can lead your journey with you."

I was embarrassed to think my journey would be a public event (and also take up so much of the group's time).

"Close your eyes…" I can't remember what Steve said exactly, but then he asked, "What do you see?"

"I am in Lidl, looking at the vegetables. A dragonfly is sitting on the potatoes, looking at me. The dragonfly is flying out of the door."

"Follow him."

"He is going down the road towards Kinvara. He's flying into the Burren Nature Sanctuary, he's in the waterfall garden. He's looking at the waterfall… He wants to go through the water."

"Tell him to go through the water."

"He has flown through the waterfall. There is a cave behind it. It is lit with sunlight. It is decorated with beautifully embroidered wall hangings, there are cushions all over the floor, and a little fire is burning in the centre. There is a round table with fruit piled on top of it. It smells of roses. When you look out the cave door, the waterfall is making a rainbow."

"This is your home."

After the workshop, I started introducing people to their totem animals. Perhaps Steve had given us all a spiritual gift that day. But I believe anyone can do this—all you have to do is ask. It can be beneficial to uncover answers from our body, or higher selves, that we don't realise we already know.

BONUS ACTIVITY: ANIMAL TOTEM HEALING

Children are especially good at this and don't even need to be relaxed. Just ask them to close their eyes. If you're working with an adult, ask him/her to lie down. Start the relaxation. Move from the toes up, asking them to relax each part of their body: feet, ankles, calves, their bones, their organs, all of it. Ask the following questions and take note of the animals that appear. The person will forget them quickly.

"Focus on the root chakra." If the person is not familiar with chakras, tell them what part of the body to focus on. If you are not familiar with chakras, make a printout to guide you.

"Say to yourself, 'What animal wishes to appear?'"

After a few seconds, they will say something.

"I can see a frog (a bird/a buffalo/a snake/a rabbit, etc.)."

"Ask the frog if it needs anything."

"It wants a pond (some food/a friend, etc.)." The answers are personal and infinitely varied. If the person doesn't speak for a while, ask them to tell you what animal they can see.

"Ask the frog if it wants to tell you something." Sometimes the response will be "no", so just move on, and sometimes there will be a message.

"Thank the frog and say goodbye."

Move up to the next chakra and continue until you have completed all seven. You can do the whole activity in ten minutes. Normally, there are some helpful, profound answers that the participant understands (but you may not). At other times, they realise the meaning later. Give them the list of animals. You are really not 'doing' anything. You are just leading the person to run the session for himself. Occasionally, a challenging problem comes up that seems to need resolving. In such a case, you can ask him to call an assembly of all the animals. Tell the person to imagine a beautiful clearing in a forest and to invite all the animals in. Tell him to ask them who wants to speak, and who they want to talk to. You can easily do animal totem healing for yourself in bed before you go to sleep. The animals are always available to help!

Jan told me about the heart-warming sessions she had done with young offenders, using Steve's techniques. At first, there was always a lot of bravado and defensive posturing. These teenagers were hard cases who may have been in gangs or convicted of violent crime or drug dealing. She would start the sessions, and after a while, someone would say something like:

"I am a growing flower. I can feel my roots going into the soil and my stem reaching up to the sky…"

At the end of the sessions, the teenagers were always calmer and happier and thanked her warmly.

To the Native Americans, animals are very symbolic. If an animal crosses your path or appears in your mind's eye, you can Google the totem meaning and see if there may be a message.

For example, if you Google the frog, you will find results like this: "The appearance of the frog spirit animal symbolises a prosperous and abundant time for you."

Have fun with your totem spirit animals!

NATURE LOVER INTERVIEW: JOSHUA STYLES

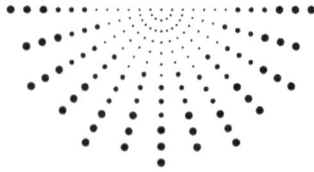

I just find plants amazing!
They are the fundamental basics of most life on earth.

— JOSHUA IS A CONSERVATION ECOLOGIST AND FOUNDER OF THE NORTHWEST RARE PLANTS
INITIATIVE IN THE UK

HOW DID YOU BECOME A NATURE LOVER?

I prefer to think of myself more as an obsessive!

I love wildlife in general, particularly plants. I started getting interested when I was six or seven years old. My mom was really kind and let me use the whole of the back garden to grow fruit and veg.

Then one year, I remember watching Monty Don on Gardener's World and he encouraged people to grow wildflowers. And so I did! I vividly remember sitting down for hours watching swarms of solitary bees and butterflies and so much stuff congregate around this teeny, tiny wildflower patch.

After that, I dug out all my fruit and veg, chucked them all in the bin and replaced everything with native wildflowers. So it started there!

WHAT IS YOUR FAVOURITE PLANT?

> I admire all plants of course, particularly all of our native species at the moment, I really love great sundew. It's a carnivorous plant, and it's just beautiful. I love it.

WHAT CAN PEOPLE DO TO HELP NATURE?

> I only mow my lawn once a year and it's turned into a superb wildflower meadow. It's swarming with wildlife. So that's one thing you could do. You could dedicate parts of, or your entire garden to wildlife.
>
> And another thing that's obvious to me, is donate to your local wildlife trust, move with your wallet, volunteer with them.
>
> Both of those things can make fantastic differences.

IF YOU HAD A MAGIC WAND, WHAT IS THE ONE THING YOU WOULD YOU DO TO HELP THE PLANET?

> It may sound a little bit severe. I'd wave my magic wand and stop people having more than two kids. It's maybe a bit controversial, but no one talks about overpopulation and we're in a massive biodiversity crisis at the moment.
>
> I would wave my magic wand to make people want to conserve things. I'd want to wave my magic wand to make people want to become more educated about the things moving around in their gardens and in their locality. And off the top of my head, I wish that there weren't endangered species in the world. There you go!

Visit www.naturemagic.ie (or search 'Nature Magic' on all major podcast platforms to listen to the full interview).

Learn more about Joshua here: http://nwrpi.weebly.com

PLANTS OF THE BURREN

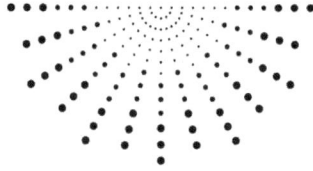

INTRODUCTION: THE MAGICAL BURREN

Walking across the warm, grey rocks of the Burren hills in early spring, it is intoxicating to discover the bright, cobalt blue dots of the first spring gentians scattered like galaxies across the grassy plateaus.

They appear before the rest of the flowers, heralding the arrival of spring and the bounty of flora about to reappear over the 100 square miles (250 square km) of the limestone hills.

Even someone who suffers from the worst case of Plant Blindness will feel that wave of excitement. It is impossible not to be drawn down, to marvel at the little piercing cobalt blue stars, their six diamond-shaped petals surrounding a white centre atop a purple stalk. Unsurprisingly, they are the symbol of the Burren.

In the words of Patrick McCormack, "Mystics will tell us that is the first colour we will see when we pass over to the other side."

Visitors find it hard to grasp what, or even where, the Burren is. Basically, it is a geographical area of karst limestone that stretches from the coastal village of Kinvara to the famous Cliffs of Moher, inland to Kilnaboy (home of the sitcom series character, Father Ted) and north towards the leaning, round tower of Kilmacduagh, and the warm, shallow waters of Lough Bunny.

The Burren is renowned for its alien landscapes of limestone pavement with eerie, deep crevices. Its most famous and magical hill is Mullaghmore, a lopsided, rocky sunken soufflé, always reflecting in the lake at its foot.

There are castles and dolmens, waterfalls running uphill; lone, windblown bonsai fairy trees and miles of stone walls and green roads. The 'light' that draws artists shines low under battleship-grey clouds on neon-green fields, turning sheep and stone walls pink. Rainbows decorate the soft, grey hills that sit reassuringly on the landscape like a clutch of broody hens.

The Burren is also the most biodiverse place in Ireland. On this enchanting backdrop, the mesmerising wildflowers steal the show with their indescribable beauty. From the weird and wonderful exotic wild orchids like the bee orchid, frog orchid, fly orchid and ladies' tresses orchid; to the swathes of thorny burnet roses, to the mats of incongruent mountain avens (an arctic plant), to the delicate, sub-tropical maidenhair fern.

Alkaline-loving plants flower next to acid-loving plants; arctic plants next to tropical plants, next to flowers that look like bees. And filling the gaps are bright pink blankets of the bloody cranesbill, a wild geranium; and splashes of the yellow and orange bird's-foot trefoil. Every botanist has the Burren on their wish list...

So what combination of random events conspired to produce this magical land?

Three-hundred-and-fifty million years ago, the Burren was a tropical sea situated on Earth where Brazil is now.

Plate tectonics have shifted millions of tons of compressed, tropical fish bones across the Atlantic to the western edge of Ireland, forming the conspicuous limestone rocks that form most of the Burren today. Perhaps this is how the sub-tropical maidenhair fern arrived.

This land was a coniferous forest until the previous ice age devastated the landscape and stripped the soil. Glaciers carried large boulders of granite and deposited them atop hills. That ice age probably brought the seeds of the mountain avens.

The forest grew back, deciduous this time, and the pockets of trees left in the Burren hark back to this time: oak, hazel, holly and ash. Humans chopped these trees down and started grazing animals; and the trees never found their dominance again.

There was no water for the farmers' herds on the hills in the summer, so they brought cattle or sheep down to the valley pastures in spring, allowing the wildflowers space to bloom. This process is called 'transhumance.' In the rest of the world, farmers put their stock up in the mountains during the summer. The Burren, as is its topsy-turvy character, does the opposite. In winter, the springs fill up and the pastures between the rocks are fertile and nutritious; and cattle can thrive.

Rainwater is slightly acidic and has sculpted the rocks into grooves and channels, and runs down through the crevices or 'grikes', forming underground rivers and extensive cave systems with astonishing calcareous stalactites.

The Burren has protected itself from the worst effects of land 'improvements' as it is pretty much impossible to turn these weird, rocky hills into regular fields. As such, they have become one of the last wild havens for biodiversity.

Peregrine falcons circle above, hares flit over miles of stone walls, pine martens stealthily search for eggs, and between the golden sands of Fanore and the Aran Islands, basking sharks sieve the water for plankton; and dolphins jump playfully.

The Burren inspired Tolkien's Middle-earth, and Poll na Gollum cave was the inspiration for Gollum.

Burren Nature Sanctuary is a beautiful introduction to this magical landscape with its unique flora and fauna. The sanctuary is on a 50-acre (20-hectare) organic site and features Burren habitats, which showcase the flora in its natural setting: shattered limestone pavement; calcareous, orchid-rich grassland, ash and hazel woodland and a unique tidal freshwater turlough.

Our dome-shaped Burren Botany Bubble is home to the Living Collection of Burren Flora. We set up this collection in 2012, and it's being developed continually to conserve and showcase the vast array of native wildflowers and orchids in the Burren Region. Visitors can identify wildflowers and orchids in the Burren Botany Bubble exhibit before heading off on nature walks around the Burren Nature Sanctuary or longer hikes in the surrounding Burren hills.

Burren Nature Sanctuary is now partnered with the National Botanic Gardens, Glasnevin, and is working on the new National Seed Bank as a satellite resource for seed collecting. See 'The Flora of the Burren' in our bonus material:

http://www.naturemagic.ie/bonusmaterial

PLANT BLINDNESS: AN EPIDEMIC

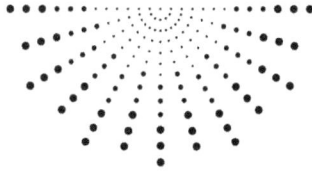

Lots of people love plants, but most people only love some plants—especially big, bright bunches of flowers. Many people suffer from what we like to call Plant Blindness. If you ignore plant species that don't shout out, "Here I am!" with bright, large splashes of colour, you are suffering from Plant Blindness.

How do you get people to connect with the plants they walk past daily?

It is a much more significant challenge to engage people with plants than with animals. But it is the next step, and just as important.

It can be pleasantly surprising to see how plants weave their magic if you give people time to discover them. Before we can address the ultimate challenge of reversing the worldwide crisis of biodiversity collapse, we must tackle Plant Blindness.

PLANT LABELLING

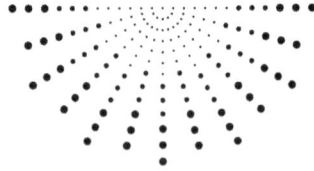

One simple step in the right direction is plant labelling.

You don't form a relationship with someone if you don't know their name. You are much more likely to say hello the next time you meet that person if you do. Likewise, if you give a plant a name, it suddenly has an identity. Then you can greet it the next time you see it and maybe introduce it to a friend.

Imagine walking past a low, bushy patch of little creeping white rock roses and being able to say:

"This is the mountain avens. It grows in the Arctic tundra, and it came to the Burren when the ice retreated 10,000 years. It tracks the movement of the sun across the sky during the day, a phenomenon called heliotropism."

Instead of:

"Look at the little white flowers," or even walking right past them…

Every plant has a name and a story to tell.

Emer Ni Dhuill is an Irish speaker and picked out a few Burren plants with interesting Irish names. She has made up a table with the Latin, Irish and English names and a column with information on the translation/folklore/medicinal uses. She has kindly shared the information on the following pages:

INFORMATION ON A NUMBER OF PLANTS OF THE BURREN

Latin Name	Irish Name	English Name	Other information
Achillea millefolium	Athair thalún	Yarrow	[8]Athair is the Irish word for 'father' and 'creeper'; talún is the Irish word for 'ground', 'land' and 'terrestrial'. [2]Widely valued for altering blood flow in a variety of beneficial ways. Known in parts of Ireland as the 'herb of the seven cures'. Chewing the leaves or smoking them in a pipe can make toothache disappear. Also used to induce nosebleed in order to relieve migraine and headache. (pg 301-302) [2]Edible green leafy vegetable (pg 67). [2]Herbal teas (pg 68). Medical use – to staunch external bleeding (pg 96). As a tonic for week blood (pg 97). For colds and chills (pg 99). Reputed to be aphrodisiacs of love potions (pg 119). [2]Folklore - Hart (1898) wrote of the following charm involving yarrow being placed under a pillow to foretell future loves 'Good morrow, good morrow, fair yarrow, / Thrice good morrow to thee; / I hope, before this time to-morrow, / You'll show my true lover to me.'
Adiantum capillus-veneris	Dúchosach	Maidenhair fern	[2]Drinks and cordials (pg 66). [3]Medical use for asthma (pg 95). [2]Staunch bleeding (pg 67). [2]Coughs and sore throats (pg 99). [2]Used as a substitute for tea (Deenan 1927) (pg 627)
Anthyllis vulneraria	Méara Muire	Kidney vetch	[8]Méara is the Irish for 'mayor' and 'fluting', but in this case it is most likely the grammatical form of 'méar' which means 'finger', plural 'méara' and Muire is 'Mary'. [2]Cuts and wounds (pg 100), [2]May have been used in the middle ages as a source of a yellow dye (pg 183). [4]Species name *vulneraria* means wound in Latin
Asperula cynanchica	Lus na haincise	Squinancy wort	[8]'haincise' is possibly a grammatical form of 'aincis' which is the Irish word for ' fussiness'. [3]Recommended as a 'cure' for quinsy or the king's evil in olden times. A virtually unknown disease now. (pg 41) [2]Treatment for quinsy – inflammation of tonsils and throat (pg 108). [2]Common from Cos Kerry to Galway by absent from the rest of Ireland (pg 199)
Asplenium ruta-muraria	Luibh na seacht ngabh	Wall rue	[5]Gábh in Irish means 'predicament' and seacht is 'seven'. [1]In Ireland it has been identified as a 'herb of the seven gifts' valued in Tipperary for its ability to cure seven diseases, and in Cavan it was boiled in milk

Emer Ní Dhúill, July 2020

Scientific name	Irish name	English name	Notes
			and taken for epilepsy (pg 62). [1]Its common and Latin names indicate that this fern thrives on walls and in rocky crevices (pg 630). [2]Wall-rue was use to cure epilepsy by taking nine pieces of cranium of a dead man and grinding them down to make a fine powder and dissolving it in a decoction of wall-rue; while fasting, the patient swallows a portion of the liquid until it is gone and if it is unfinished, the owner of the skull will return to look for the unswallowed portion (pg 630)
Asplenium trichomanes	Lus na seilge	Maidenhair spleenwort	[5]Seige is the Irish word for 'hunted' and 'hunting'. [2]Cameron (1883) said the name 'sealg' means 'spleen' in Scots Gaelic and that is was formerly used as a remedy of ailments associated with the spleen (but no ethnobotanical uses have been noted in Ireland) (pg 631). 5 Spleen in Irish is 'splín'.
Campanula rotundifolia	Méaracán gorm	Harebell	[5]Méaracán is the Irish word for 'thimble' – Méaracán gorm translates to 'blue thimble'. [2]Harebell is the official flower of Antrim (County Flowers) (pg 132 and 223). [2]Common in the North and West, but much rarer elsewhere in Ireland (pg 223).
Cirsium dissectum	Feochadá móna	Meadow thistle, Bog thistle	[1]*Cirsium* spp. thistles. Known in Limerick to staunch bleeding in cattle or horses after lancing. (pg 351). [2]*C. dissectum* found in west, north-west and centre of Ireland and rarely distributed elsewhere (pg 258). [6]Ireland holds >25% of the European population and is categorised as LC on Irelands Red List of Vascular Plants (pg 30). [4]Thistles are the larval foodplants of the Painted Lady Butterfly.
Corylus avellana	Coll	Hazel	[1]Ash from burnt hazel stick was put on burns in Monaghan; bark was applied to boils in Kerry. (pg 89) [2]One of Ireland's most important native plant resources – used for food, timber, building, basketry, fuel, medicine and many other purposes (pg 265); hazelnuts were an extremely important food source in ancient Ireland and provided valuable carbohydrates and fats (pg 265); there is reference to early law texts in Old Irish that mention that a hungry person was entitled to gather a handful of hazelnuts from privately owned land (pg 265); folklore – forked hazel twigs (gabhlog in Irish) were used by dowsers or diviners for locating underground water (pg 269). [3]Ash from burnt hazel stick was put on burns in Monaghan; bark was applied to boils in Kerry. (pg 89)
Cymbalaria muralis	Buaflíon balla	Ivy-leaved toadflax	[8]Buaflíon translates to 'toad-flax' and balla is the Irish for 'wall'. [2]Noted by Phillips (1983) as an edible herb in salads – but no records of traditional Irish use (pg 276)

Erinus alpinus	Méirín sí	Fairy foxglove	[5] Méirín is the Irish for 'finger'. Méirín sí translates to 'fairy fingers'. [4] Fairy foxglove grows on old stone walls and is not very commonly found in Ireland.
Eupatorium cannabinum	Cnáib uisce	Hemp-agrimony	[5] Cnáib is Irish for cannabis, hemp, marijuana; uisce is the Irish for 'water'. [2] There are no ethnobotanical uses noted for Ireland; it is a frequent plant throughout Ireland of river and stream banks, marshes and rocky places (pg 305). [4] Much loved by butterflies and bees.
Filipendula vulgaris	Lus braonach	Dropwort	[8] Braonach' is the Irish word for 'dewy' or 'dripping'. [4] Dropwort spreads by means of slim thread-like root that carry starchy tubers at their ends; the Latin '*filum*' means 'filament' and '*pendulus*' means pendulous – the species name 'Filipendula' likely refers to these roots. [2] Possible archaeophyte; very rare in Ireland – found in limestone grassland only in a few square kilometres of the eastern part of the Burren in Cos Clare and Galway; as medicinal plant – used in treatment of epilepsy, kidney and bladder stones, genital discharges and intestinal worms (pg 318).
Gentiana verna	Ceadharlach bealtaine	Spring gentian	[5] Ceadharlach is the Irish word for 'gentian' and bealtaine is 'spring'. [2] Widespread in west of Ireland from north Co Clare to south Co Mayo; particularly abundant in the Burren (pg 330). [6] Assessed as Near Threatened in Irelands Red Data List (pg 28). [2] Featured on an Irish postage stamp in 1978 (8p) and 2004-2008 (€10) (pg 137 and 138).
Minuartia verna	Gaineamhlus earraigh	Spring sandwort	[5] Gaineamh is the Irish work for 'sand', lus is 'herb and earraigh is 'spring'. [2] Found in rocks and screes and some sand dunes in only a few places Ireland in Co. Clare, the Aran Islands, Antrim and Derry (pg 405)
Ophrys apifera	Magairlín na mbeacha	Bee orchid	[2] Featured on an Irish postage stamp in 1993(28p) and 2004-2008 (7c) ((pg 137 and 138); rare but widespread in Ireland on limestone grasslands, dry banks and sand dunes (pg 420). [6] Categorised as LC in Ireland's Red Data List for vascular plants (pg 86). [7] In European folklore, the orchid was seen as a symbol of fertility and an aphrodisiac. [7] The name Orchid comes from the Greek *orkhis* which means testicle, which is the same meaning as the Irish name *magairle* (The tubers being compared to testicles). So Magairlín na mbeacha could be directly (and incorrectly) translated to Bees Testicles! – But really it translates to Bee Orchid (i.e. 'magairlín' is the Irish word for 'orchid' which is a derivative of the word 'magairle' which is 'testicle' (plural = magairlí).

Pinguicula grandiflora	Leith uisce	Large-flowered butterwort	[8] Leith is the Irish word for 'flounder' and uisce is 'water'. [4]Native carnivorous plant. [2]Very common in boggy habitats in parts the of south-west, such as Co Kerry and West Cork. [2]Reported that butterwort leaves used to curdle milk; Williams (1993) reported that it was used for rennet in Ulster (pg 439). [2]Was reported from 1726 to be used to treat swellings in the 'clefts' in cow's udders; Moloney (1919) said that the Irish name *léith uisce* denotes the disease of liver fluke in sheep (pg 440) (I could not find any translation for the word léith).
Potentilla fruticosa	Tor cúigmhéarach	Shrubby cuinquefoil	[5]*Tor* is the Irish work for a *bush* or *shrub*. 'Cúigmhéarach' means five fingers – referring to the 5 petals I presume. [6]Categorised as VU in Ireland's Red Data List (pg 91). [2]Rare native shrub found on lake-shores and rocky habitats that are regularly flooded, such a turloughs.
Primula veris	Bainne bó bleachtáin	Cowslip	[8]Translates to '*Juice of cowslip*'. [2]An occasional plant in the central plain of Ireland of old unimproved grasslands and especially roadsides; much rarer in the NE and SW of Ireland (pg 456); said to be a narcotic and was used to treat insomnia (pg 456)
Saxifraga hypnoides	Mórán caonaigh	Mossy saxifrage	[8]Mórán is the Irish for 'many' and 'much' and caonaigh is the Irish for 'moss'. [2]Rare except in the Burren; elsewhere, this species is a boreal montane element in the European flora (pg 520)
Senecio jacobaea	Buachalán buí	Ragwort	[8]Buachalán is the Irish word for 'ragwort' and buí is 'yellow. [1]Considered to be used for jaundice in many parts of Ireland. Locally popular for colds, coughs and sore throats. Used for cuts, sores and inflammation for various kinds. Applied to burns or scalds and poulticing of boils or abscesses. Also valued for rheumatic complaints, sprains or swollen joints. Also used for measles, dropsy, bowel hives in children, warts and nettle stings. (pg 307). [2]If eaten this plant is poisonous to cattle and horses although harmless to sheep; declared a noxious weed under the 1937 Order of the Control of Noxious Weeds Act (1936) (pg 532). [2]Used quite extensively in Ireland as a medicinal plant; leaves applied to animal wounds as a poultice; used for humans in the same way for treatment of cancers (pg 533)
Spiranthes spiralis	Cúilín Muire	Autumn lady's tresses	[8]Cúilín is the Irish word for 'point' or 'treadle'. Muire refers to Mary. [6]Categorised as NT in Ireland's Red Data List (pg 105)
Succisa pratensis	Odhrach bhallach	Devil's-bit scabious	[5]ballach is the Irish work for 'spotted', 'blotchy' and 'macular'. I couldn't find a translation for 'odhrach'. [1]Used for sores and boils. (pg 278). [2]Apparently it received its common name because it contained so many cures that it 'angered the devil', so he cut its root short (pg

		561). [4]This species is the main larval foodplant of the Marsh Fritillary butterfly.

REFERENCES

1. Allen, D. E. and Hatfield, G. (2004) *Medicinal plants in folk tradition: An ethnobotany of Britain and Ireland*. Timber Press, Cambridge.
2. Jackson, P.W.J (2014) *Ireland's generous nature: The past and present uses of wild plants in Ireland*. Missouri Botanical Garden Press.
3. Nelson, C (1999) *Wild plants of the Burren and Aran Islands*. The Collins Press
4. Wildflowers of Ireland (http://www.wildflowersofireland.net/)
5. Foclóir.ie – new online English-Irish dictionary (https://www.focloir.ie/)
6. Wyse Jackson, M., FitzPatrick, U., Cole, E., Jebb, M., McFerran, D., Sheehy Skeffington, M. & Wright, M. (2016) *Ireland Red List No. 10: Vascular Plants*. National Parks and Wildlife Service, Department of Arts, Heritage, Regional, Rural and Gaeltacht Affairs, Dublin, Ireland.
7. Mac Coitir, N. (2015) *Ireland's Wild Plants Myths Legends and Folklore*. The Collins Press.
8. Teaglann.ie online dictionary and language library (https://www.teanglann.ie/)

At the sanctuary, we build on our labelling every year and have Latin, English and Irish names on our botanical labels. The more we label, the more plants seem to cry out to have their name badges! Recently, during the Covid-19 lockdown period, there was a guerrilla botanist in Kinvara who labelled the weeds along the pavement with chalk, so the children could learn their names on their daily 1,2 mile (2 km) walk around the village: 'nettle', 'ragwort', 'dock'—all important plants in their own ways.

Another easy way to spark joy around plants is, as the relentless activist Jonathon Porritt suggested, to grow plants from seed. Grow a few lettuces or some flowers, even if it's in a window box.

Gonçalo Santos, another Nature Magic podcast guest and amateur botanist, tells his story of how a visit to the garden centre (to buy a potted mint plant for Mother's Day) led to him creating a large, communal garden from wasteland around his apartment block; and to a lifelong passion for plants. Visit www.naturemagic.ie (or search 'Nature Magic' on all major podcast platforms to listen to the full interview).

An excellent activity for children is to fill a glass with damp cotton wool and put a sunflower seed between the glass and the cotton wool. Watch the magic as the roots sprout down and the cotyledon (the first leaves to develop) uncurls and reaches up. Then transfer your seedling to a large pot, and your child will marvel at the colossal flower growing out of the tiny seed.

See 'The Parts of the Plant' on our printable poster in the bonus material: http://www.naturemagic.ie/bonusmaterial

When my eldest was about five years old, I peeled a section of lawn away, and she scattered some wildflower seeds. I didn't have much hope, but we were happily awe-struck! I have a photo of her sitting in her tiny, round meadow: a burst of colours with poppies and cornflowers. She never forgot it.

You can easily follow Mary Reynolds' advice and give half of your garden back to nature, "Become part of the solution and not of the problem," is her advice.

If given a chance, our plant friends will return. To learn more about giving a part of your garden back to nature, follow Mary's project at http://wearetheark.org.

A WORD ABOUT FARMING

Modern chemical farming considers weeds as its number one enemy. Conventional, non-organic farmers are at constant war with 'weeds'.

But these weeds are all valuable in their own right—even ragwort, the torment of every horse owner. It is the food source for the cinnabar moth. The much-vilified nettle, that we are continually slashing and pulling, is a vital food source for butterflies and contains important qualities for herbal remedies.

Until we respect our biodiversity and understand that the web of life supports us as humans, we are going down a dangerous and ugly road. (See the Web of Life Activity on page 202.)

Organic farming does not use pesticides or sprays that wipe out biodiversity. It uses a system of rotation and low-stocking density to work with the land and allows nature to thrive and support the soil.

Our farming systems must change. Listen to local farmer Patrick McCormack on the Nature Magic podcast at www.naturemagic.ie (or search 'Nature Magic' on all major podcast platforms to listen to the full interview).

The Burren Nature Sanctuary site has a chemical-free life history, and the 50-acre (20-hectare) site has had official organic status with the Irish Organic Association for 25 years.

Go organic!

EXAMPLE ACTIVITY: GROW A NATIVE WILDFLOWER MEADOW

This is easy to do as long as you follow what we do with our 15-acre calcareous (calcium-rich), orchid-rich meadow at Burren Nature Sanctuary.

Our farm had a head start as it has been certified as organic for 25 years. We have never used chemical fertilisers or pesticides (of any variety) on it. However, even a conventionally farmed meadow or lawn will revert to a meadow over time if managed correctly.

Don't worry if, initially, you find a lot of nettles and docks. If you keep following the process, you will have a beautiful, diverse and colourful meadow after a few years.

Follow these simple steps:

- Hay (or bales of haulage/silage, depending on rain) is cut in late September and removed from the field. This helps to provide the nutrient-poor environment that wildflowers love. In a garden, mow over during the winter and remove the clippings. Wildflowers thrive in a nutrient poor soil because there are no fast-growing tough grasses to compete with that might overwhelm them in richer soil.
- The meadow is grazed until 1 February. We have cattle and horses and every couple of years, borrow some sheep who do a really tight trim! This is a balancing act as you want them to strip as much of the grass as possible, and you must not feed them on the meadow as this would replace nutrients (which does not suit the wildflowers).
- All livestock are taken off the meadow from 1 February until the end of September, allowing the flowers to bloom and produce seed.

That's it! The biodiversity of our meadow is astounding. Some common wildflowers that grow at the sanctuary are dog daisies, yellow rattle, clover, bird's-foot trefoil and a range of wild orchids.

The meadow is at its finest in June, July and August. Don't buy packets of wildflower seed. Instead, allow the native species space and time to return.

It can be hard to sit and meditate, but plants can make this healthy practice simple. The following activities will help you become more engaged with nature and become a calmer, better you.

Identify Wildflowers

Take a wildflower book and a magnifying glass on your walk. Tick off the flowers you see along the road. Take your time to study their individuality. Notice what colours they are, how many sepals and petals they have, what season they flower in and how many there are in the population.

Learn to recognise them, so you can greet them by name in the future.

Picking Fruit Meditation

Focus your mind on the parts of the flower: the pollen, the petals, the stamen. Think of the process the fruit goes through during development.

The petals fall as summer turns to autumn, and the fruit ripens.

Think of the vitamins the fruits are giving you; and about the animals that eat them to disperse the seeds and continue the cycle of life.

In spring the bee drinks the flower's nectar, brings pollen to the ovary, and leaves with fresh pollen from the stamens of the flower.

Tending a Garden

As you are weeding your vegetables, ask for your negative thoughts to be removed, allowing space for positive thoughts to prosper.

WORLD BULLYING DAY: AN IMPORTANT LESSON ABOUT NURTURING PLANTS

In 2018, a branch of IKEA in the Middle East set up an experiment for World Day of Bullying Prevention.

IKEA placed two plants in a school lobby and put up signs for each plant. One plant's sign said, 'this plant is being complimented' while the other plant's sign said, 'this plant is being bullied'. They played audio files with positive messages to the one plant and negative messages to the other plant for 30 days. The children joined in and also recorded positive and negative messages, which were replayed to the plants.

After 30 days, the plant that was being bullied was slowly withering away while the other plant (that had been receiving positive messages) was thriving.

This is a sad but powerful and insightful experiment that illustrates the intelligence of plants.

To learn more about how plants can respond to human care and nurturing, read The Secret Life of Plants by Peter Tompkins and Christopher Bird.

BRACKEN: A BANDAGE FOR THE LAND

Speaking of bullying plants, I had a bad relationship with bracken for a long time… It popped up everywhere, taking over the paths and fields, quickly covering the flowers with its destructive pointy growth in spring.

Apparently, its spores are carcinogenic. All I could ever think about when it came to bracken was, "How are we going to get rid of it?" Mow it, let the goats eat it away, pull it out…

The only upside I had ever heard was that people ate the young shoots as they unfurled in spring, and they tasted like Asparagus. I never quite believed this and was always too wary of being poisoned to try it!

Then, one day, I heard a naturalist call bracken 'the bandage of the land'. I learned that when soil is exposed, nature needs to cover it as quickly as possible. Bracken instantly does the job, holding and protecting the ground until young bushes and (eventually) trees take over the responsibility. I now have a new and profound respect for bracken.

I apologise for all my negative thoughts against it!

There may be certain things you simply don't like in this world or your personal life. But that doesn't mean they are worthless. Sometimes, the things we don't like or understand may exist for a deeper, meaningful reason.

FELLOW NATURE LOVER INTERVIEW:
DR NOELEEN SMYTH

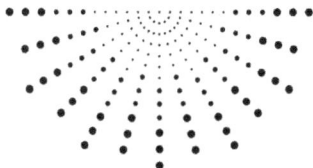

As humans, we think we are the masters of our domain. But really, we're just sharing this place. We wouldn't be here without plants.
Life wouldn't exist without them.

— DR NOELEEN SMYTH IS THE CONSERVATION OFFICER AT THE NATIONAL BOTANIC GARDENS
IN GLASNEVIN, DUBLIN

HOW DID YOU BECOME A NATURE LOVER?

I was lucky to have had a rural childhood in County Kildare, surrounded by boglands, which I absolutely adore; it is just a lovely habitat.

WHAT IS YOUR FAVOURITE PLANT OR ANIMAL?

Well, one plant which really fascinated me when I was a kid was Arum maculatum, which is also known as 'lords and ladies'.

Every kid in Ireland is told to, "STAY AWAY! STAY AWAY! THEY ARE POISONOUS!"

You see those luminous orange berries in autumn and you are instantly drawn to them. But they are more fascinating in the springtime if you look inside. They have a sinister hood, called a spade, and inside is this one little spike, like a finger. It's called a spadix. So that gets a bit warmer and emits a foul odour and attracts little flies from everywhere. Then, when they crawl in and they're looking for the source of this delicious stinking scent, they crawl down past all these hairs, which point just in one direction, and then they are down into a little chamber. They are there to fertilise and pollinate the plants, they can't crawl back up as the hairs keep them trapped down there, they are one-directional.

So when we opened it up, you could see all the little flies that were trapped by this Arum until they got the job done. They get trapped in their little prison, pollinating away with no reward!

I love the tricks plants play and these hidden things that we don't often see.

DO YOU FEEL SPIRITUALLY CONNECTED WITH NATURE?

I think Ireland is a very magical and spiritual country, anyway. We live surrounded by all these Celtic things, like down in Clare you have the dolmens, the landscape is dotted with them; we are immersed in it.

I'm looking at a map. There seems like there isn't a square kilometre where there isn't a holy well or a holy cross or a holy something; it's a very ancient landscape.

There are some very magical places. Glennstall Abbey in Limerick is an amazing place. The Clare Glens are down around there—a wonderful green oasis. I've been following the Killarney fern around all these magical elfin places. They are very special to me.

If you had a magic wand, what is the one thing you would you do to help the planet?

66 I suppose the tropical rainforest is just my instant wish—put those right back. These are the lungs of the Earth to replenish our oxygen.

Visit www.naturemagic.ie (or search 'Nature Magic' on all major podcast platforms to listen to the full interview).

THE BURREN BOTANY BUBBLE
NATIVE PLANT COLLECTION AND EXHIBIT

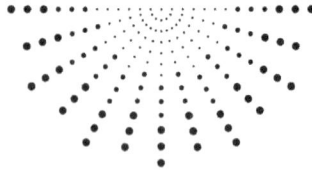

The Burren Botany Bubble is an eye-catching, dome-shaped polycarbonate structure. Exhibiting native plants is quite a challenge. Wildflowers are small and fleeting. Also, ensuring the right balance of ventilation and water in a closed environment is tricky.

Interpretation of the flora is key. Explaining to people what they are looking at in a fun, engaging, but educational way is a delicate/tricky balance. See our flora interpretive panels in the bonus material:

http://www.naturemagic.ie/bonusmaterial

The collection of Burren species in the Burren Botany Bubble mostly come from the farm, with some donated specimens originating elsewhere.

You require an exclusive license to move plants listed in the Flora Protection Order. http://www.irishstatutebook.ie/eli/2015/si/356/made/en/print

To build on the collection, it is necessary to apply for a license.

Under Section 21 of the Wildlife Act, an Order was made in 2015, entitled Flora (Protection) Order, declaring individual plants to be protected throughout the State. Under Section 21 it is an offence for a person to cut, pick, uproot or otherwise take, purchase, sell or be in possession of any plant whether whole or part, of a species mentioned in the Order, or wilfully to alter, damage, destroy or interfere with the habitat of such a species, except under licence of the Minister, and then strictly for Scientific, Educational or other such purposes.

Through Botanic Garden Conservation International, we are partnered with Jardí Botànic de Sóller in Spain. They hold the botanic collection of the Balearic Islands in Majorca, and have a long-established native plant nursery and seed bank, and repopulate areas that have lost biodiversity.

I will discuss this partnership in more detail below, under the heading 'The Wild Orchid Project'.

Ultimately, we aim to accomplish the same goal with our living collection. We are partners with the National Botanic Gardens in Glasnevin, Dublin and we're working together on the new National Seed Bank.

CONSTRUCTION OF THE BURREN BOTANY BUBBLE

We have both an outside garden and an inside exhibit. Our exhibit is a geodesic dome which was constructed by a local polytunnel company.

We have a rainwater drip watering system. We collect rain from the donkeys' shed roof, and gravity feeds it to the Bubble. (The public water supply can become salty and was not suiting the native plants.)

This year, we replaced the electric thermostat fan with a lifted, vented roof for better circulation and to allow more heat out on hot days. We also have six vents of 2 ft x 2 ft (60 cm x 60 cm) at ground level. The two double doors are left open year-round (unless there is a significant weather event).

Maintaining the collection has been a seven-year learning curve. The primary management for the plants is reducing grasses that can swamp the vulnerable native plants like orchids or gentians. In the wild, the grass would be eaten away every winter. To mimic these conditions, we have to remove the grass manually.

We are lucky to have a supremely talented plantsman, Edward Dee. Each native plant requires its own unique combination of circumstances: it may like the sun, the shade, to be damp, to be dry, to get frost, to not get frost, etc. The plantsman must be a magician to maintain the correct habitat for each plant.

- People like to pick flowers, especially children. This was illustrated very clearly with the abundance of flowers in the Botany Bubble during the COVID-19 lockdown, when we had no visitors for four months, over springtime.
- Initially, we had put up signs saying: 'Please do not pick the flowers, this is a botanical collection'. But these didn't seem to work. We replaced them with an image of a hand and a flower with a red cross over it. Even if you can't read, it conveys the message!

A SPECIAL THANKS TO SUSAN SEX

When we opened the Burren Nature Sanctuary, we were very honoured to be gifted with botanical prints and a set of Irish postal stamp designs, "Irish Wildflowers", by the supremely talented botanical artist, Susan Sex, author of Ireland's Wild Orchids: A Field Guide. Thank you, Susan. It was a very kind and supportive gesture.

THE WILD ORCHID PROJECT

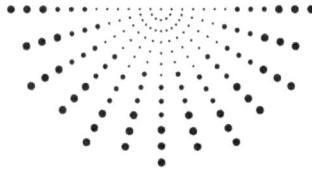

Since opening to the public in 2012, Burren Nature Sanctuary has been a member of Botanic Gardens Conservation International (BGCI). Their expertise and support have been invaluable in managing and developing our collection of native Burren flora.

In 2018, BGCI chose us to partner with Jardí Botanic de Sóller, the National Botanic Garden of Mallorca. We were grant-aided by BGCI to travel to each site. In October 2018 we went to Mallorca, and in July 2019 staff from Mallorca travelled to the Burren.

We received a visit from Jardí Botanic de Sóller staff in July 2019. The tour was arranged to progress the Joint Wild Orchid Project, and so they could learn about our education initiatives.

Together, we visited the National Botanic Gardens of Ireland in Glasnevin and presented a bee orchid seed to the conservation officer, Dr Noeleen Smyth.

During October 2018, staff members from our sanctuary visited the Sóller Botanic Garden. We received invaluable training in seed collection, banking, native plant collections and much more.

Going forward, Burren Nature Sanctuary will be a satellite resource for Burren native plants for the National Seed Bank.

We would like to thank Botanic Gardens Conservation International, The National Botanic Gardens of Ireland and Jardí Botanic de Sollér for supporting this conservation initiative.

THE SÓLLER BOTANIC GARDEN

The Sóller Botanic Garden is a private foundation devoted to the conservation of the flora of the Balearic Islands. It was founded in 1985 and opened to the public in 1992. It holds the collection of wild plants from the Balearic Islands and, in their seed bank, conserves more than 60% of the endemic, rare and endangered plant species from the Balearic Islands.

They designed their living plant collection in a landscape of terraces with rocky garden features. Besides the primary collection comprising the Balearic Islands' flora, they also house plant collections from the Canary Islands and other Mediterranean islands.

They also keep a collection of medicinal plants and other small groups of cactuses, succulents, carnivorous, and other ornamental species which can live in the Mediterranean.

PARTNERSHIP ACTIVITIES

Between the Burren and Mallorca, there is a common list of wild orchids. In the Sóller Botanic Garden, most of the wild orchids of the Balearic Islands are cultivated in their living plant collection.

We will build on the wild orchid collection at the Burren Nature Sanctuary by leaning on their experience and employing the help of Sóller Botanic Garden staff.

Within our partnership, the primary activities will be:

- Inventory and geo-referencing of the orchids' populations living in the Burren Nature Sanctuary.
- Inventory of other plants growing in the same area to define the habitat for orchids.
- Search for a suitable site to extend our living orchid collection.
- Adapting the landscape for the orchid collection.
- Collecting seeds to store in a seed bank.
- Translocation of some orchids to start the collection and establish the habitat.
- Labelling the plants for public visitor information.

- The proposed partnership activities support the goals of Botanic Gardens Conservation International:
- "To mobilise botanic gardens and engage partners in securing plant diversity for the well-being of people and the planet."
- Sóller Botanic Garden has started implementing a few things they have learned from our educational program during their visit:
- Adaptation of the Leave No Trace School Tour in Sóller town.
- Start a project that uses an audio guide app.
- Design a botanical walk.

On the next page you'll find the 'Orchids of the Burren' table. Orchids that grow in both the Burren and Mallorca are highlighted.

Latin Name	English Name	Irish Name	Rarity
Anacamptis pyramidalis	Pyramidal orchid	Magairlín na stuaice	Common
Cephalanthera longifolia	Narrow-leaved helleborine	Cuaichín caol	Rare
Coeloglossum viride	Frog orchid	Magairlín an loscáin	Frequent
Dactylorhiza incarnata ssp. cruenta	Leopard orchid	Magairlín craorag	Local
Dactylorhiza fuchsii subsp. fuchsii	Spotted-orchid	Nuacht bhallach	Abundant
Dactylorhiza fuchsii var. okellyi	O'Kelly's spotted orchid	Nuacht bhallach Uí Ceallaigh	Frequent
Dactylorhiza maculata	Heath-spotted orchid	Na circíní	Abundant
Dactylorhiza incarnata ssp. pulchella	Early marsh orchid	Magairlín álainn	Frequent
Dactylorhiza majalis	Broad-leaved, western marsh orchid	Magairlín gaelach	Rare
Epipactis atrorubens	Dark-red helleborine	Cuaichín dearg	Frequent
Epipactis helleborine	Broad-leaved helleborine	Ealabairín	Frequent
Epipactis palustris	Marsh helleborine	Cuaichín corraigh	Local
Gymnadenia conopsea	Fragrant orchid	Lus taghla	Frequent
Listera ovata	Common twayblade	Dédhuilleog	Abundant
Listera cordata	Lesser twayblade	Dédhuilleog bheag	Very rare
Ophrys apifera	Bee orchid	Magairlín na mbeach	Local
Ophrys insectifera	Fly orchid	Magairlín na gcuileanna	Local
Orchis mascula	Early purple orchid	Magairlín meidhreach	Abundant
Anacamptis morio	Green winged orchid	Magairlín feltbeach	Very rare
Neotinea maculata	Irish orchid, dense flowered orchid	Magairlín glas	Local
Neottia nidus-avis	Bird's nest orchid	Magairlín neide éin	Rare
Platanthera bifolia	Lesser butterfly orchid	Magairlín beag an fhéileacáin	Rare
Platanthera chlorantha	Greater butterfly orchid	Magairlín mór an fhéileacáin	Rare
Spiranthes spiralis	Autumn ladies tresses	Cúilín Muire	Rare

ST THÉRÈSE OF LISIEUX, "THE LITTLE FLOWER"

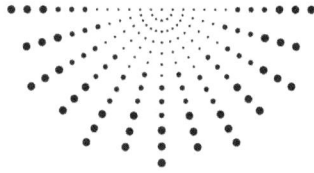

My great grandfather, Edward Hearne, fought in the trenches in WW1.

He was in the veterinary core and ended up with the sad task of butchering injured army horses to feed the troops. It must have been a desperate job. He always kept a prayer card of St Thérèse of Lisieux with him. When they were under attack, he used to pray to her; and it is said that the card curled up in his hands. After the war, he had his first child and named her Thérèse, in honour of the saint. I have inherited the name and the prayer card.

St Thérèse is still a much-beloved saint and a familiar presence in Irish houses nowadays. Her nickname is "The Little Flower".

Below are some of her famous quotes. She died when she was only 14.

> I understood that every flower created by Him is beautiful, that the brilliance of the rose and the whiteness of the lily do not lessen the perfume of the violet or the sweet simplicity of the daisy. I understood that if all the lowly flowers wished to be roses, nature would no longer be enamelled with lovely hues. And so it is in the world of souls, our Lord's living garden.

> When I am in heaven, I will let down a shower of roses.

— ST THÉRÈSE OF LISIEUX (BORN JANUARY 2, 1873, ALENÇON, FRANCE—DIED SEPTEMBER 30, 1897, LISIEUX; CANONISED MAY 17, 1925; FEAST DAY OCTOBER 1)

NATURE LOVER INTERVIEW: MARY REYNOLDS

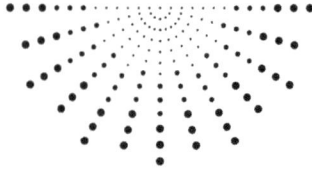

My vision with the ARK project is to create a patchwork of healed little pieces of heart that we can wrap around the earth.

— MARY REYNOLDS IS A NATURE ACTIVIST AND
THE YOUNGEST EVER GOLD MEDAL WINNER AT THE CHELSEA FLOWER SHOW.

HOW DID YOU BECOME A NATURE LOVER?

I grew up on a farm in Wexford and I had a really strong relationship with 'my green family' when I was a kid.

I did a garden in Chelsea which took off into an international career in garden design, and it was all about wild places but still very controlled.

The winter before last I was in my office, looking down over my lawn. My landlord let me re-wild half of the property, about an acre, all brambles, etc.

A fox ran past, then a couple of hares, then a family of hedgehogs, and they were supposed to be asleep. So, I thought, 'okay, there's something going on here'. At the end of my lane someone had gotten planning permission and there was a thicket of many years of growth, a scrubby, thorny woodland. It was a home for so many creatures. They got a digger and cleared it all out to make a lawn. I stood there and realised I had done this so many times, and that was the end of my garden design career.

I went home and thought about the fact that these creatures have nowhere left to go because we keep tidying everything up, thinking we are doing something good. Farming has turned into such an attack on nature, there is nowhere safe there.

So, I started We Are the ARK, 'Acts of Restorative Kindness' to the earth and I am encouraging people to let go of the concept of gardening.

DO YOU FEEL SPIRITUALLY CONNECTED TO NATURE?

The experience that influenced my work most was when I was a kid, there were six of us, we were farmers and I had a lot of freedom. I wandered off to a field at the top of the farm. It was May, and the hawthorns were flowering. I remember the smell; I was five or six. The gap in the field closed behind me, it sounds crazy, but I knew something was happening behind me and the 4-foot gap in the field behind me was gone. I ran around the field and eventually got distracted by the sun and the butterflies in the meadow. It was the late 70s and the green revolution hadn't kicked in, so there wouldn't be many chemicals. I sat down and felt surrounded, and I felt the plants had spirits and personalities like people.

WHAT CAN PEOPLE DO TO HELP NATURE?

If people give their land back to nature, build an ARK, something magical happens. Instead of being gardeners, they will become guardians and carers for all sorts of wild creatures that will come to live in their ARKS.

It opens the heart to share our piece of this earth with as many creatures as possible, to support them in their time of need. This simple act causes people to fall back in love with nature, and once you fall, there's no getting out.

IF YOU HAD A MAGIC WAND, WHAT IS THE ONE THING YOU WOULD YOU DO TO HELP THE PLANET?

I would ask people to give half of the Earth back to nature, including the seas, then we have a future.

Start with our own bits. Even if we give half of our gardens back to nature, we are part of the solution and not of the problem.

Visit www.naturemagic.ie (or search 'Nature Magic' on all major podcast platforms to listen to the full interview).

Learn more about Mary and her initiative, We are the ARK, at http://marymary.ie and check out her new book, The Garden Awakening: Designs to Nurture Our Land and Ourselves.

HERBAL APOTHECARY GARDEN & HERBAL REMEDIES

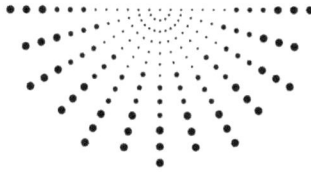

"Apothecary" refers to a medical professional who formulates and dispenses medicine. Early apothecaries' investigations of herbal and chemical ingredients were the precursor to modern pharmacology sciences. In recent times, this term has mostly been replaced by the word 'pharmacist'.

Our Herbal Apothecary Garden helps us connect people to their natural surroundings in a fun and engaging way. It also offers a simple way to interpret native plants.

If you have a nature centre or you work at one, consider doing something similar with the herbs in your area.

HOW WE CREATED THE GARDEN

We collected common native herbs that grow abundantly in fields, hedgerows and on the side of roads and transplanted them to our decorative herb bed.

Edward Dee, our wonderful plantsman, constructed an appealing Celtic serpent-shaped herb bed with different compartments. Each common herb has a botanical label, so our visitors will quickly be able to identify these common plants if they walk past them in their daily lives. Adjacent to the herb bed, we have informative Herbal Traditional Remedy Interpretation Panels. Each panel has images and information about the most relevant plants. They also display simple recipes and techniques to make use of the herbs' medicinal properties.

See 'The Herb Panels' in our bonus material: https://www.naturemagic.ie/bonusmaterial

HERBAL REMEDIES

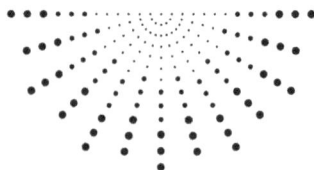

HERBS FOR COUGHS AND COLDS

Thyme: Antiseptic and expectorant syrup. Add 1 lb. of honey to 600 ml of strained infusion and take 1 tsp as required.

Elder Flower: Antiviral and anti-inflammatory; boosts immune system. Cordial (see recipe below).

Agrimony: Antiviral, healing. Gargle 1 cup of cool tea as required for sore throats.

Eyebright: Anti-inflammatory soothing. Use cooled infusion as an eye bath or compress.

Mullein: For ear infections. Use a couple of drops of infused/macerated oil as required.

Wild Garlic: A general antibiotic. Make wild garlic pesto.

RECIPE

Elderflower Cordial (for an immune boost)

Ingredients:

- 1 kg white sugar,
- 2 litres water,
- 2 unwaxed lemons.
- 20 fresh elderflower heads, 50 g citric acid (from the chemist).

Method:

- Dissolve sugar and water in a pan over low heat and then bring to boil.
- Allow to cool.
- Pare and slice lemons and add along with citric acid. Wash the flowers carefully in a large basin and add. Stir well.
- Cover the pan and leave to infuse for 24 hrs.
- Line a colander with a clean tea towel, sit over a large bowl or pan.
- Ladle in the syrup and strain.
- Use a funnel and a ladle to fill sterilised bottles.
- Will keep in the fridge for up to 6 weeks.

HOW TO MAKE

An Infused/Macerated Oil

Ingredients:

- 100 g dried herbs or 300 g fresh herbs.
- 500 ml vegetable oil (olive/sunflower).

Method:

- Place the finely chopped herbs in a heatproof bowl and add the oil to cover.
- Place over a pan of boiling water and heat gently for 2 hours, top-up as necessary.
- Strain and repeat using new herbs.
- Strain and pour into a sterilised dark glass bottle, name and date.
- With fresh herbs, let oil stand before bottling to allow water to sink to the bottom, pour off oil.
- Store in a cool place and use within 3 months.

HERBS FOR MIND AND EMOTION

Oats: Antidepressant, restorative nerve tonic. Eat daily (see recipe below).

St John's Wort: Relieves mild depression, SAD, nervous exhaustion, menopausal moods. Make a tincture; take 1-4 ml twice daily. Consult a doctor before use. (May cause photosensitivity.)

Vervain: Relieves insomnia, slight sedative. Make a tea infusion from the aerial parts of the plant, drink 1 cup before bed.

RECIPE

Oat Porridge - for a steady release of energy

Method:

- Put the organic oats in a saucepan, pour in the milk or water and sprinkle in a pinch of salt.
- Bring to the boil and simmer for 4-5 minutes, stirring from time to time and observing that it doesn't stick to the bottom of the pan.

HOW TO MAKE

A Tincture

Ingredients:

- 200 g dried/fresh herbs (dry fresh herbs to remove moisture).
- 1 litre of 37.5% proof vodka.

Method:

- Chop the herbs finely and place in a large, sterilised sealable jar.
- Immerse the herbs.
- Seal the jar and store for 2 weeks away from direct sunlight, shake occasionally.
- Strain through a muslin cloth and filter through an unbleached coffee filter.
- Pour into a sterilised dark glass bottle; label with name and date.
- Use within 12 months.

Rose: Grief and mild shock. Make a syrup: dissolve 8 oz sugar, juice of 1 lemon & 1 orange slowly at low heat. Add 100 rose petals.

Use as cordial or over pancakes.

Mugwort: Clears negative stuck energy, calms and protects. Make a smudge stick.

Pick when flowers are open and dry for a few days make a bundle 3 cm thick and bind in a spiral with cotton thread.

HERBS FOR DIGESTION

Wood Sage: Improves digestive function. Take one cup of infused leaves 3 times daily.

Mint: Digestive tonic relieves nausea. Make a tea infusion and drink as required.

Plantain: Relaxing expectorant and healing to mucous membranes. Treats gastritis and IBS. Make a tincture and take 60 drops 3 times daily.

Juniper Berries: Carminative (reduces flatulence and discomfort), stimulates digestion. Take 20-40 drops of tincture 3 times daily.

Bramble leaves: Treats diarrhoea and supports general health. Make a strong tea infusion from bramble leaves and drink to treat diarrhoea.

Dandelion: Improves liver function and digestion. Make a salad from leaves (see recipe below).

RECIPE

Dandelion & Primrose Leaf Salad for a Spring Detox

Ingredients:

- 30 g dandelion leaves,
- 1 tsp chopped wild garlic leaves,
- 10 g yarrow leaves,
- 20 g primrose leaves,
- 10 g rocket leaves,
- 1 head of chicory,

- 1 1/2 tbsp linseed oil,
- 1 1/2 tbsp lemon juice, toasted sesame seeds,
- White pepper and salt to taste.

Method:

- Wash leaves well and dry in a spinner. Toss with seeds and add the dressing.

HOW TO MAKE

An Infusion / Tea

Ingredients: 1 heaped tsp dried herb or 2 tsp chopped fresh herbs, 175 ml boiling water.

Method:

- Place the herbs in a cafetière or teapot, steep for 10 minutes. Can be drunk cold throughout the day.

HERBS FOR HEART & CIRCULATION

Wild Garlic: Improves circulation. Add leaves and flowers to salads.

Dandelion: Stimulates lymphatic system. Make a tea infusion from dried flowers.

Nettle: Tonic, circulatory stimulant, diuretic, anti-allergenic. Drink 1 cup of tea infusion 3 times daily.

Hawthorn Berries: Peripheral vasodilator, cardiac tonic, relaxant. Take 2 cups of decoction of berries daily. (Do not take alongside other heart medication.)

RECIPE

Garden Greens Juice (for a general pick me up and heart tonic)

Ingredients:

- 2 large dandelion leaves,

- 6/8 young nettle leaves,
- A handful of hawthorn berries,
- 4 handfuls of kale leaves/chard/spinach leaves,
- 1 cucumber,
- 2 stems of fresh marjoram,
- Half of an unwaxed lemon.

Method:

- Wash and juice.

HOW TO MAKE

A Decoction (used for woodier parts of a plant)

Ingredients: 15 g of dried herb or 30g of fresh herbs.

Method:

- Place chopped herbs in a saucepan and pour in the water.
- Cover the pan and bring to the boil, then simmer gently for 15-20 minutes.
- Strain and divide into 3 doses for use that day.

HERBS FOR FIRST AID

Yarrow: For cuts. Use ointment on wounds and grazes, for nosebleed stuff the nostril with a leaf.

Red Clover: For Stings. Crush the flowers directly on to stings.

Self Heal: For Wounds. Use oil on cuts, sores and wounds. Use fresh leaves on clean wounds. Tea infusion can be used for flu and fevers for its antiviral, antibacterial and cooling properties.

Cleavers: Grazes, inflammations, blisters and rashes. Apply a compress soaked in tea.

Rose: Grief, shock. Take as a cordial or drizzled over pancakes.

Eyebright: Sore eyes. Bathe in cooled tea made from an infusion.

RECIPE

Rose Petal Syrup

Ingredients:

- 225 g granulated syrup,
- juice of 1 lemon,
- juice of 1 orange,
- 100 g of dried rose petals.

Method:

- Dissolve the sugar in 300 ml of water over low heat. Do not boil.
- Add the strained lemon and orange juices. Simmer on low heat for 5 minutes.
- Over 15 minutes add the rose petals allow to cool and strain.
- Keep refrigerated and use within 6 weeks.

HOW TO MAKE

A Solar Infused Oil & A Herbal Ointment/Cream

Ingredients: (for oil)

- 300 ml of infused oil,
- 25 g beeswax.

Ingredients: (for cream)

- 300 ml of infused oil,
- 25 g beeswax.

Method:

- Make a solar infused oil by covering the yarrow leaves with olive oil.

- Cover with a cloth and leave on a sunny windowsill for two weeks, stirring occasionally.
- Strain and pour off, leaving any water droplets behind.
- Warm the ointment ingredients in a double boiler (a bowl above a saucepan of water)
- Pour into jars and label.
- If making cream, add the warm infusion slowly while beating on the slowest speed.

HERBS FOR MUSCLES & JOINTS

Comfrey: Soothes pain of broken bones, relieves blind infections. Apply a cool compress to unbroken skin.

Yellow dock: Rheumatism, osteoarthritis. Take tincture 20-40 drops 3 times a day. Ground-up roots are a gluten-free flour used in famine times.

Burdock: Bruises, rheumatism, arthritis, gout. Apply poultice.

Horsetail: Rheumatic pains, sprains, chilblains. Make a strong tea infusion and add to bath.

Chickweed: Rheumatism, shingles. For a chickweed bath, put a few handfuls in a sock and hang under the hot tap. Add oatmeal for additional soothing.

RECIPE

Chickweed Pesto

Method:

- Pick a few handfuls, removing brown bits and roots.
- Break off the larger stems, put the rest in a blender with a handful of pine nuts and add a couple of cloves of garlic and enough oil to make it bendable.
- Serve with grated parmesan, eat with pasta, rice or as a sauce for vegetables.

HOW TO MAKE

A Poultice

Method:

- Steam the herbs to soften, wrap in a soft cloth and apply as hot as possible.
- Keep reheating.
- Use a hot water bottle on top of poultice and keep on as long as possible.

Meadowsweet: Arthritis, rheumatism, neuralgia. Apply cream/ointment/balm.

People love to take photos of the signs and investigate the herb bed to see what the plants look like in 'real-life'. Then, when they go on our nature walk, or even walk in their own garden or local park, they quickly identify the herbs they learned about at the herb bed.

We only include common herbs that would easily grow in most people's gardens in our herb bed. Also, we never encourage people to forage for wild, native plants (instead, they should raise their own herbs). We follow the Leave No Trace principle, which is to leave whatever you find in its natural environment.

NATURE LOVER INTERVIEW: JONATHAN PORRITT

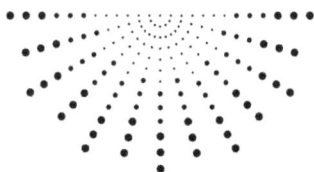

If I could really wave that magic wand, one thing that would matter most to me would be that every single child in the world could be brought up in such a way that his or her contact with nature could be sustained.

— JONATHAN PORRITT IS AN ENVIRONMENTAL ACTIVIST

HOW DID YOU BECOME A NATURE LOVER?

I was lucky; I spent quite a bit of time living in New Zealand when I was 18 to 20 and one of the things I did there was to plant more than 200,000 trees.

There was a particularly special bit of land there on three acres called The Cathedral. This little patch of land was just quite amazing, it was the only patch that was proper native bush. I spent so many hours planting trees on this place and then always taking refuge in The Cathedral when the rain became too difficult to deal with or it was too cold. You have to plant trees in the winter, obviously.

It was amazing. I spent as much time on the land then as I've ever spent throughout the rest of my life. So that was really what got me into it.

WHAT IS YOUR FAVOURITE PLANT OR ANIMAL?

The slime mould is my favourite species! I think they are quite astonishing creatures.

There are hundreds and hundreds of different species of slime mould all over the world. We have learned the critical importance of slime moulds to the whole fragile way in which the rainforest works.

They are quite special.

They are little single-celled organisms, and during the autumn they come alive. They move into the active phase and they form these great multi-celled bits of glob. If you see them on the forest floor they are not beautiful, but they are critical to the wellbeing of the forests. They eat a lot of the stuff, the detritus that's left over. And then even more amazingly, once they've eaten everything, they have some astonishing mechanism that turns them from these single celled fungal organisms into spores. These spores are blown everywhere else in the woodland.

DO YOU FEEL SPIRITUALLY CONNECTED TO NATURE?

A spiritual connection with nature has been a really important part of my work as an activist. When things got tough over the last 45 years, I've often used my love of nature as a way of putting things to rights in my own mind and carrying out some short-term, temporary healing.

Just to look, to understand, to observe much more closely, sometimes to be silent, to think about the beauty of nature around us.

I do think one of the most astonishing things going on in the Covid-19 crisis is the number of people now who are commenting on birdsong. I cannot imagine in any other crisis in human history, where birdsong would suddenly be featuring so prominently! Just give yourself space to be in nature and enjoy it for what it is.

Or grow stuff! For kids, the opportunity to plant something themselves, see it grow, pick it and then eat it, can be a memorable experience. Even if it's just some herbs in a window box!

IF YOU HAD A MAGIC WAND, WHAT IS THE ONE THING YOU WOULD YOU DO TO HELP THE PLANET?

Well, the technical answer is that I would make sure that every government in the world introduced a carbon price. Until we make people pay for the emissions of these greenhouse gases that are causing the climate emergency, we're never going to get an intelligent, viable economy.

So we have to have an economic system that allows governments to impose a price on carbon. Every tonne of CO_2 and other greenhouse gases that are emitted would carry a certain price. And that money could then be used to fund the restoration of nature to help deal with a lot of today's social problems.

A little bit more realistically, I guess if I could really wave that magic wand, one thing that would matter most to me would be that every single child in the world could be brought up in such a way that his or her contact with nature could be sustained.

Visit www.naturemagic.ie (or search 'Nature Magic' on all major podcast platforms to listen to the full interview).

Link to Friends of the Earth: https://www.foe.ie

Link to Survival International: https://www.survivalinternational.org

TREES

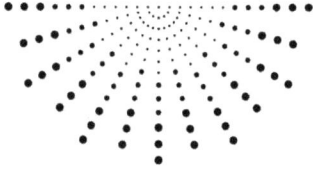

THE IRISH SCOTS PINE: A TALE OF PERSISTENCE

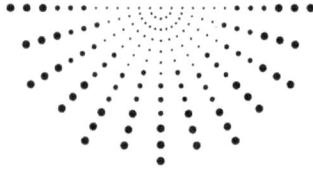

Until recently, it was believed that the native Irish Scots pine (*Pinus sylvestris L.*) had become extinct. This is thought to have happened during a massive population decline about 2,000 years ago throughout Europe.

The Scottish variety had survived and were used to replant the trees where they had been lost, including Ireland.

However, a new study of a small population in Rockforest (County Clare, Ireland) has proven that they are native to Ireland. A paper by Alwynne H. McGeever and Fraser J. G. Mitchel titled Re-defining the natural range of Scots Pine: a newly discovered microrefugium in western Ireland, shows how fossil pollen from soil cores was radiocarbon-dated to prove this particular population is the original, native Irish Scots pine!

Our wonderful plantsman, Edward Dee, sourced three saplings from the native population in Rockforest in 2019. They are now happily growing next to the nature walk between the hawthorn arbour and the Wishing Hut on the turlough (our Disappearing Lake) side of the path.

HOW THE IRISH SCOTS PINE SURVIVED

The Rockforest population was found growing in an area of limestone pavement that is unattractive to convert into agricultural land. They were within the bounds of an estate which would have afforded some protection to the woodland.

Seeds were collected under license by the National Parks and Wildlife Service. Each tree now has a colour-coded label with a link to a photograph of the parent tree.

THE MAGIC OF LEAVES

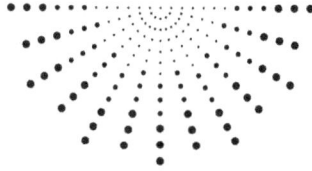

As every schoolchild knows, forests are the lungs of the Earth.

The great forests capture sunlight with their leaves and provide us with oxygen. They capture the carbon and use it to build their skeletons.

For 400 million years (400,000,000), the world's ancient tropical forests in the Congo and the Amazon, and the Northern Boreal snow forests, have been the carbon banks of the Earth.

In the last 100 years, humans have devastated these forests.

HOW LEAVES WORK THEIR MAGIC

Leaves allow plants to 'drink' the sun, and they pull carbon into their bodies to form trunks, branches and leaves. Here is a simplified explanation of the process:

- Water enters the plant (generally from the roots).
- Carbon dioxide enters through little holes called the stomata.
- The short, continuous sine wave energy of sunlight bounces on the green material in the leaf called the chloroplast.
- The chloroplast grabs a unit of energy from the electrons and changes the form of the water and carbon dioxide molecules, breaking off carbon from the carbon dioxide in the process.
- This liberates oxygen into the air and pumps carbon into the body of the tree.

Ta-Da! Magic…

The genome of a tree is only different from a human by two 'bases'. In ancient wisdom, they knew that trees were our friends. We find it hard to form a relationship with these beings because they move at such a slow pace that we cannot see it and think they are just lifeless statues. But if you watch a time-lapse of a tree, moving its leaves throughout the day to follow the sun and growing over time, it is clear that they are living and moving and responding to stimuli—just at a different pace than us.

In ancient Ireland, the trees were greatly respected and protected by the Brehon Law. There were four classes of tree:

- Airig fedo ('nobles of the wood') which included oak, hazel, holly, yew, ash, Scots pine and crab-apple trees.
- Aithig fedo ('commoners of the wood') which included alder, willow, hawthorn, rowan, birch, elm and cherry trees.
- Fodla fedo ('lower divisions of the wood') included blackthorn, elder, spindle, whitebeam, aspen and juniper trees.
- Losa fedo ('bushes of the woods') included bracken, bog myrtle, furze, wild rose, bramble and heather trees.

They graded penalties for harming any of those trees according to the class of tree. For example, if someone had felled a 'noble of the wood', he would be expected to pay two-and-a-half milk cows, and if he had felled a 'commoner of the wood', he would be expected to pay one milk cow; and so on. See 'The Ogham Language' in our bonus material:

http://www.naturemagic.ie/bonusmaterial

LOWER YOUR CARBON FOOTPRINT WITH OUR NATIVE TREE PROJECT

Contribute to our native tree planting project to lower your carbon footprint.

With a small contribution of €10.00, trees will be purchased and planted at the Burren Nature Sanctuary in early spring.

Thank you!

Each tree absorbs approximately 44 pounds (20 kg) of carbon dioxide each year, that is approximately 992 pounds (450 kg) per lifetime. A thousand trees would offset the lifetime carbon footprint of the average European citizen. Visit The Burren Nature Sanctuary website:

https://www.burrennaturesanctuary.ie/product/donate-e10/

EXAMPLE ACTIVITY: FOREST BATHING, A 1400-YEAR-OLD TECHNIQUE FROM JAPAN

- Walk in a pine forest in warm weather for 20 minutes. Take off as many clothes as possible!
- Stand up straight and inhale into the bottom of your lungs.
- Pines produce a bolus of aerosols, little kite-shaped particles called alpha and beta-pinene. These anti-carcinogenic compounds, already in use in modern medicine, are adsorbed into the lungs and interact with the body.

Research from the Nippon Medical School in Tokyo (the oldest medical university in Japan) shows that alpha and beta-pinene molecules act as an alert system to the T cell ration in the body (the immune system), pressing the 'on' function. Basically, it means the 'gobbling' cells in the body increase in volume, and alertness increases for a period of up to a month, removing any cancer cells in their path.

Learn more about forest bathing by watching To Speak for the Trees by Diana Beresford-Kroeger: https://www.youtube.com/watch?v=IBOVet8Ki4g

THE SONG OF WANDERING AENGUS, A POEM BY W.B. YEATS (1865–1939)

I went out to the hazel wood,
Because a fire was in my head,
And cut and peeled a hazel wand,
And hooked a berry to a thread;
And when white moths were on the wing,
And moth-like stars were flickering out,
I dropped the berry in a stream
And caught a little silver trout.

When I had laid it on the floor
I went to blow the fire a-flame,
But something rustled on the floor,
And someone called me by my name:
It had become a glimmering girl
With apple blossom in her hair
Who called me by my name and ran
And faded through the brightening air.

Though I am old with wandering
Through hollow lands and hilly lands,
I will find out where she has gone,
And kiss her lips and take her hands;
And walk among long dappled grass,
And pluck till time and times are done,
The silver apples of the moon,
The golden apples of the sun.

NATURE LOVER INTERVIEW: EDWARD DEE

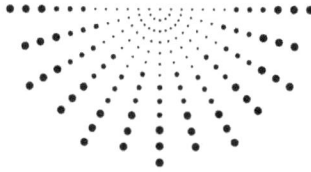

I have had a huge hunger for the natural world since I was a boy; it seems to get more and more absorbing as time goes on.

— EDWARD DEE IS THE PLANTSMAN AT BURREN NATURE SANCTUARY
AND FOUNDER OF THE LIVING COLLECTION OF BURREN FLORA.

HOW DID YOU BECOME A NATURE LOVER?

I seem to have been born with an innate curiosity for the natural world. As a boy, I knew every bird's nest around the farm. I learnt their calls and habits. My sister pressed wildflowers and kept them in a book and named them. She kept me interested with her knowledge and quiet passion. There was a lot more wildlife in general at that time.

I am lucky to have spent my youth in the Yorkshire countryside.

WHAT IS YOUR FAVOURITE ANIMAL OR ANIMAL?

My favourite animal is the badger. We have to stop persecuting them!

My favourite plant is blue bells, carpets of them, the smell is intoxicating at full tilt.

My favourite rock is sandstone embedded with quartz. I have a lovely piece in my garden, crystal points and all.

WHAT CAN PEOPLE DO TO HELP NATURE?

Join the Irish Wildlife Trust (https://iwt.ie) and be active locally.

Join the Woodland League and get kids involved in the Forest in a Box Project (http://www.woodlandleague.org/forest-in-a-box-project/) Kids plant a box with native tree seeds. One year later, they plant them on and hopefully make the connection and get the interest and passion to make a difference.

Start a small nursery—Green Friends Ireland (https://greenfriendsireland.wordpress.com) will take your sapplings and plant them on.

Join Flora and Fauna International (https://www.fauna-flora.org).

Read the full blog post on our website: https://www.burrennaturesanctuary.ie/burren-nature-sanctuary-blog/

MAGICAL STORY
MOTHER MAPLE

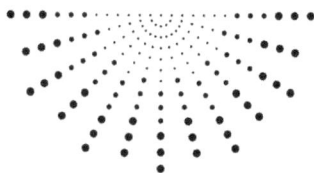

In the summer of 2018, I took my 13-year-old daughter to upstate New York. She desperately needed treatment for previously undiagnosed, late-stage Lyme disease. We were there for eleven weeks while she attended a clinic for IV antibiotics. The antibiotics cause the severe Jarisch–Herxheimer reaction (when harmful bacteria die their toxins are released into a patient's bloodstream, causing a painful inflammatory response). In the fourth week, she hit rock bottom. They stopped the medication because her body couldn't tolerate it. She was in agony and unable to get out of bed.

My mother came over to help for a week, and I was able to leave the apartment to take a walk in the nearby Cohoes National Park. I sat down on a picnic bench, put on my earphones and listened to Imelda May singing Black Tears, and sobbed. A bright blue dragonfly sat on the table and stared at me. A hummingbird flew past, squirrels came out to play, a groundhog popped out of the bushes, and blue jays landed on branches nearby. It was like a scene out of Snow White.

I looked into the depths of the forest and appealed to the trees to help Lorna. I vowed to dedicate my life to nature. This book is part of my thanks.

The following night, I dreamt I was standing very close to a tree and looking at the trunk. It said, "My name is Mother Maple, and I am going to help you." I woke up and checked the 'Trees of New York' guide, and sure enough, the bark I had seen in my dream was maple.

On our way to the clinic, we visited the house of the myofascial masseuse therapist who worked at the Stram Center. She introduced us to Abi, her

19-year-old friend who had also suffered severe Lyme disease before she recovered with Lisa's help. Lisa wanted Lorna to keep the image of Abi as a beacon of hope. I walked around the side of the house. A large tree stood in the middle of the back garden — it was the tree from my dream.

Like an excited child, I exclaimed: "I know the name of your tree, its Mother Maple!"

Lisa got Lorna through the terrifying treatment with myofascial massages and energy healing sessions. She was an integral part of her journey to recovery. Today, I know that the message from the maple tree was that Lisa was the person that would help the situation. She is still a dear friend and has visited us in Ireland.

Thank you, Lisa and Mother Maple. Lisa sends me photos of her tree in its seasonal apparel: spring leaves, summer leaves, autumn leaves and a snowy winter coat.

The sugar maple is the official state tree of New York.

A Note on Ticks

At least 30% of ticks carry Lyme disease and other infections, and they are widespread across Europe and the USA. Avoid long grass and use a natural tick repellent.

If you get bitten by a tick, save it in a zip-lock bag and freeze it so you can get it tested. Consult your doctor and see www.ticktalkireland.com for advice on treatment.

NATURE LOVER INTERVIEW: LAURA POWERS

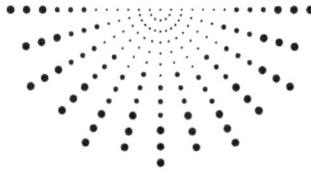

Fairies are the guardian angels of the plant and animal kingdoms.

— LAURA POWERS IS A PODCAST HOST, CELEBRITY PHYSIC, AND COACH

HOW DID YOU BECOME A NATURE LOVER?

I've always loved plants and gardens in particular. I have always been very drawn to animals. I was a horse lover as a girl. I loved animals of all kinds, and so I think I've always felt connected.

And then, as I developed my psychic abilities, fairies started to show up. So, I was really drawn in very quickly into that magical aspect of nature as well.

WHAT IS YOUR FAVOURITE PLANT OR ANIMAL?

Trees are amazing, they're so wise and very healing. You can lean up against a tree and just feel the strength that you receive. They have a consciousness and an intelligence, and they do so much for us. In addition to providing shade and shelter, they give us oxygen; we make things from wood and we can use leaves for many different things. Flowers are also amazing; we literally wouldn't have food without the flowers.

I have always been connected with cats in particular. I love them very much. And actually, we have past lives; and I was a cat.

WHAT CAN PEOPLE DO TO HELP NATURE?

Making choices of using renewable resources and getting solar panels or things like that, those are relatively easy things for us to do that can make a huge impact on our carbon footprint and pollution in the air.

Also, eating and consuming things that are created locally, because that automatically has a smaller carbon impact; because there is less travel and fuel and you are supporting your local economy as well.

IF YOU HAD A MAGIC WAND, WHAT IS THE ONE THING YOU WOULD YOU DO TO HELP THE PLANET?

End pollution, and that includes air pollution and trash. The amount of trash that's generated every day by the average person is astounding and also, I think, can be changed fairly easily with some minor adjustments with how we operate.

And then the other one would be for people to appreciate and respect nature more. I think so often we disregard it. And then also some people almost feel like they have to conquer it or something, instead of understanding that we are a part of it. And it's, as you know, a collaboration. And in particular with plants, I find that a lot of people just completely discount plants and their intelligence. In the past, we had a lot more respect for plants and what they can provide for us.

AN EXTRA QUESTION FOR LAURA! CAN YOU TELL US ABOUT FAIRIES?

> Fairies are incredibly magical. They can be in this realm, but they can also go into their own realms.
>
> Fairies and humans used to be in the same realm all the time. But humans started attacking fairies and fairy-realm beings.
>
> They retreated into their own realm and they can go back and forth, they're not here all the time like it used to be. They like to connect with people who love and respect the natural realm, plants and animals, and also who usually have a life purpose or mission or doing something for the natural realm as well.

Visit www.naturemagic.ie (or search 'Nature Magic' on all major podcast platforms to listen to the full interview).

FAIRIES

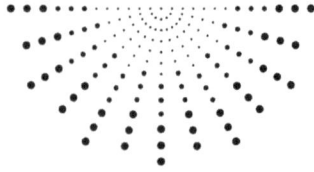

FINDING FAIRIES

Irish townlands are small, geographical divisions within parishes (districts). There might be a few different townlands around one small village. The townland where the Burren Nature Sanctuary is located is Cloonasee. It means 'Meadow of the Fairies'. There is a belief that a supernatural race, the Sídhe (shee) or 'the fairy people', used to live alongside humans. But after being persecuted, they now live out of sight in a different dimension.

Young children often see fairies as they stroll around the magical hazel woodland at Burren Nature Sanctuary. Using the concept of fairies to teach children about caring for nature is ideal. As guardians, they protect the animals, plants and trees.

Fairy Trees

The hawthorn tree (commonly called 'whitethorn' in Ireland) is thought to be a sacred meeting place for the fairies.

In Ireland, people avoid cutting down a lone hawthorn tree at all costs. These 'fairy trees' are believed to bring luck and prosperity to the landowners. Many people who visit the fairy trees leave prayers, gifts and personal items as a token of goodwill, hoping to receive good fortune or healing in return from the 'wee folk'.

The hawthorn blooms at the time of the festival of Beltane, on 1 May. This was a significant time for the ancient Irish and the Sídhe.

It is bad luck to cut down a hawthorn tree. The route of the Ennis motorway was delayed for ten years (and eventually diverted) to avoid removing a fairy tree.

Fairy Forts

The Irish Heritage Council states that there are over 60,000 Fairy Forts (also known as ringforts) In Ireland. Although most are probably iron age forts, there is a strong tradition that fairies have taken refuge in these stone circles, and it is bad luck to disturb them.

In fact, it is widely believed that the demise of Sean Quinn's cement and insurance empire in Cavan, and John DeLorean's sports car industry in Belfast, were directly attributable to the conscious destruction of 'sacred' Fairy Forts.

If you want to learn about first-hand encounters with fairies, borrow W.B. Yeats' book, Fairy and Folk Tales of Ireland, from your local library. It is a thick book that documents many fairy encounters!

Douglas Ross Hyde (the first president of Ireland), W.B. Yeats and Samuel Beckett all believed in fairies. The belief in fairies is still alive and well among Irish people.

The Fairy Wishing Well

When the fairies first moved into the woodland, our visitors used to make wishes and leave coins outside their doors.

Everyone is always trying to make wishes!

The coins were scattered everywhere. They were a hazard for wildlife, and, every now and then, an opportunistic child would scoop them up and fill their pockets...

My friend, Helen, is a part-time carer for children with disabilities. She came up with the idea to build a wishing well. Roy and David constructed a beautiful stone well (with a secure wire mesh over the top). Every year, the fairies receive up to €1000, which we give to a local child with special needs.

We dedicated the well to Che, a little boy who used to love visiting the Fairy Woodland. On his small stone plaque is his favourite saying: *Star star teach me how to shine, Close your eyes, Count to three, In a world of pure imagination.*

FAIRY FUN

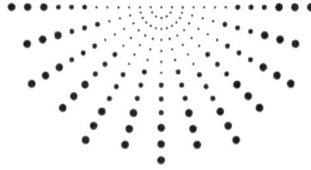

THE FAIRY TALE DRESS-UP WALK

On Friday and Saturday mornings, we host the Fairy Walk. It is immensely popular with kids under ten years old.

We have about twenty fairy dresses and dragon outfits. Half an hour before the walk starts, we wheel the costume railing out, and the kids excitedly dress up. Before we got the dragon costumes, we had knight costumes for the boys. We soon realised they were a big mistake as it encouraged the boys to engage in swordplay with sticks all the time. We find that fire-breathing dragons are a much safer option!

It is truly a beautiful sight to see twenty little fairies and dragons set off on a walk to the Fairy Woodland.

The key to maintaining control over a rowdy group of fairies and dragons is giving them simple instructions before the walk starts:

"I am your leader. Anyone who tries to pass by me while we walk must go to the back!"

Being sent to the back is the ultimate punishment for any child. Of course, they forget the instruction as soon as I give it… But if a child runs ahead, I simply freeze until they realise everyone has stopped. They quickly fall in behind me again, and the walk continues. (I never actually send anyone to the back!) When we get to the entrance of the Fairy Woodland, I wait for everyone to catch up and then tell them I need their help.

"The fairies are asleep. They work in the night and sleep in the day, so we don't want to wake them up."

Sometimes, a child may ask, "What happens if we wake them up?"

"They can get very grumpy and put knots in your hair and stuff!"

If there are rowdy, older boys intent on waking up the fairies, I have two tactics:

• Firstly, I tell them that the King Fairy is really cranky and if you wake him up, he will be furious and put nettles in your bed. They love that!

• If that doesn't work, I will take the boy aside and give him a job, or ask him to help me monitor the younger kids, or to help me with the animals later. Most kids just want attention, so this usually works. Older boys mostly want to tell you they don't believe in fairies, but by the end of the walk, they are asking all about them!

"When we go into the Fairy Woodland, I will tell you who lives in the houses. But we need to whisper. The adults usually forget, so can you please help me? They like to chat about this and that, *'Oh, I need to go to the shop to buy dinner, Mary needs new shoes, I'm going to a party tomorrow... Blah blah blah,'* so, if you hear them talking we need to shush them, okay? Can we practise?"

I shush towards the adults, and the kids follow with a little 'shush'.

"You need to be way more assertive than that."

I shush loudly at the adults, "Come on, try again."

"SHUSH."

The kids then turn around and shush the adults loudly. They love this because usually, everyone is shushing them. And besides, when do you get permission to be cross with an adult?!

Every time an adult talks from then on, we shush them, and the kids get that great sense of payback! It also makes it clear to the adults that this is for the children, not for them, and they must respect the children by not jabbering. It makes them pay attention to the children and everyone enjoys the walk even more. I stop at the first door and tell them all about the fairies. At each door, I ask a question:

"Put your hand up if you know who lives here?"

"Put your hand up if you know why the fairies need the fairy dust?"

"Anyone know where the tooth fairy gets the coins?" And so on…

Fairy doors can get the shyest child to open up. Ar Tor, the Fitness Fairy, moved in recently and they love the question, "Anyone do any sports?"

All the hands go up, "I do hurling," "I do soccer…"

Sonny, the Weather Fairy, is also a great conversation starter. "What sort of weather is it today? What other sorts of weather are there?"

The Rainbow Fairy painted her door in the Pride flag colours, which always makes same-sex couples happy. And even the youngest child can answer the questions like, "What is your favourite colour of the rainbow?" Or, "What colour does the Rainbow Fairy paint the grass, the sky (etc.)?"

By the end of the walk, they know that Zarara, the Tooth Fairy, collects the baby teeth. She gives them to King Dagda who grinds them up and hands them over to Queen Maeve who, along with 99 other ingredients, makes the Fairy Dust. And why do the fairies need Fairy Dust? To fly and do magic, of course. And where do they get the coins for the children? Obvious: Lora, the Wishing Well Fairy, dives into the Wishing Well in her little green wetsuit and gets the coins… It all makes sense!

If you want to run a similar activity, make sure you know who lives in your Fairy Houses. We get lots of repeat fairy walkers, and they won't forget!

To find out more about the fairies and what they get up to at the sanctuary, get the book, DUST: An Irish Fairy Tale from the Burren Nature Sanctuary shop. www.burrennaturesanctuary.ie

THE FAIRY POST BOX

We installed a Fairy Post Box in the woodland with a little sign saying, 'Fairies have Wi-Fi and will reply if you leave your email address'.

Children loved this. They bombarded us with some very touching (and sometimes, heart-rending) messages. Here are a few examples:

"Can you say hello to my Nana in heaven? And tell her we miss her."

"Can I be a dinosaur?"

Mostly, they were lovely drawings and questions like:

"How old is Zarara the Tooth Fairy?"

"My fairy is called Isabella, what is yours called?"

Of course, answering all the questions was hard work, and the fairies had to set up an email account, but it was worth it as children were so, so delighted to receive replies!

THE FAIRY TALE FAIRY TRAIL

Keeping children happy and engaged on a mile-long (1,6 km) trail can be tricky. It is especially hard to persuade small children to go for walks when they want to keep playing on the slide.

One day, I got an idea from my cousin, Elizabeth. She came up with a brilliant strategy after struggling to get her son to go on walks. She started telling him stories. But she'd only tell him a little bit of the story and then say, "At that tree over there, I will tell you what happened next."

Genius!

Based on what she did, we set up a Fairy Tale Fairy Trail. It starts at the Wishing Well in the Fairy Woodland. Each of our story's paragraphs is divided into little illustrated signs which can be read as you walk through the trail.

You can use the Fairy Tale Fairy Trail story if you are in Ireland or come up with a similar story to suit your native environment. See 'The Fairy Tale Fairy Trail' in our bonus material:

http://www.naturemagic.ie/bonusmaterial

INSTALLING FAIRY DOORS

Outdoor fairy doors must be made from solid wood. Most commercial fairy doors are not strong enough to withstand weather and will not survive outside. Instead, find a local carpenter who will make a sturdy door.

Try to keep the materials as natural as possible; wood needs upkeep and painting but it gives the right 'magical feel'. Touch up the doors with paint in springtime. Doors do not open and any handles, etc. must be secured so tightly that the strongest ten-year-old boy cannot pull it off!

Put a backing piece of timber against your tree and secure the fairy door to the backing piece. This is a difficult job. People are always trying to pull open the fairy doors. I recommend getting name plaques for your fairies, with their jobs written underneath. For example, 'Zarara the Tooth Fairy'. This is also useful for people who want to enjoy self-guided walks.

At Burren Nature Sanctuary, we sell wooden Burren Fairy Doors and Burren Fairy Windows for indoor use. They come with a double-sided sticker, so parents don't have to spend extra time running around, looking for a method of putting them up. The back of the Fairy Door packaging reads:

Install your fairy door in your room and leave a tiny welcome note for the fairies. If the letter is gone in the morning, you will know your fairy has moved in. Keep your room tidy as these sprites don't like bumping into things! You can ask you fairy their name or any other questions you may have.

If a child is unhappy or being bullied, the fairies can be very useful. Children can write notes to the fairies and receive replies. This, of course, entails parents writing out a letter in miniature writing and leaving it by the door in the night. Fairies always answer with compassion and love and never tell children off. But they do encourage them to keep their rooms tidy…

FAIRY RULES & FAIRY JOBS

In the back of DUST: An Irish Fairy Tale (mentioned above), we have Fairy Rules and Fairy Jobs.

Fairy Rules include:

1. Fairies MUST work at night and sleep in the day.

2. Fairies MUST not be seen by humans (without special permission from the King Fairy).
3. Fairies MUST love and respect the Animal Kingdom
4. Fairies MUST love and respect the Plant Kingdom
5. Fairies MUST protect the planet that they live on—water, air, soil and rock.
6. Fairies MUST always be kind and help each other.
7. Fairies MUST move out to their own house when they reach 100.
8. Fairies MUST be trained and have jobs by age 100.
9. Fairies may make friends with Humans in dreams or as pen pals.
10. Fairies may make friends with good Elves.
11. Fairies May make friends with sensible Pixies.
12. Fairies may make friends with happy Gnomes.

The Fairy Rules and Fairy Jobs are aimed at helping children understand the importance of taking care of the natural environment and each other. We must teach our children to love and respect nature from a young age.

THE STOLEN CHILD, A POEM BY W.B. YEATS (1865–1939)

Where dips the rocky highland
Of Sleuth Wood in the lake,
There lies a leafy island
Where flapping herons wake
The drowsy water rats;
There we've hid our faery vats,
Full of berrys
And of reddest stolen cherries.
Come away, O human child!
To the waters and the wild
With a faery, hand in hand,
For the world's more full of weeping than you can understand.

Where the wave of moonlight glosses
The dim gray sands with light,
Far off by furthest Rosses
We foot it all the night,
Weaving olden dances
Mingling hands and mingling glances
Till the moon has taken flight;
To and fro we leap
And chase the frothy bubbles,
While the world is full of troubles
And anxious in its sleep.
Come away, O human child!
To the waters and the wild
With a faery, hand in hand,
For the world's more full of weeping than you can understand.

Where the wandering water gushes
From the hills above Glen-Car,
In pools among the rushes
That scarce could bathe a star,

We seek for slumbering trout
And whispering in their ears
Give them unquiet dreams;
Leaning softly out
From ferns that drop their tears
Over the young streams.
Come away, O human child!
To the waters and the wild
With a faery, hand in hand,
For the world's more full of weeping than you can understand.

Away with us he's going,
The solemn-eyed:
He'll hear no more the lowing
Of the calves on the warm hillside
Or the kettle on the hob
Sing peace into his breast,
Or see the brown mice bob
Round and round the oatmeal chest.
For he comes, the human child,
To the waters and the wild
With a faery, hand in hand,
For the world's more full of weeping than he can understand.

MAGICAL STORY
FAIRIES LOVE HURLING

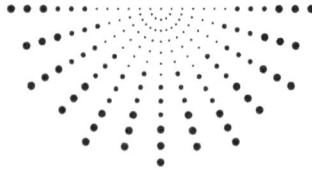

Hurling is Ireland's national sport. It is over 2,000 years old and originated in ancient Gaelic Ireland. It is played with a stick (made from ash wood) called a hurl. The aim is for opponents to hit the leather ball, called a sliotar, between the goalposts. It is the fastest field sport in the world, quite dangerous and requires a high degree of skill.

The All Ireland Hurling final is the most exciting 70 minutes of the year.

Fairies also love hurling. It is said they cannot play a game of hurling unless there is one human on the team. There are stories of people being enchanted to join a match with the fairies.

On 3 November 2019, the Kinvara Intermediate Hurling Team defeated the Kilconieron Hurling Club on a score-line of 1-10 to 0-12 and won the Galway Intermediate Hurling Championship. With two minutes to play, Kinvara was two points down.

If a player hurls the ball over the goalpost's crossbar, he scores one point for his team. If he shoots it underneath the crossbar, into the net guarded by the goalkeeper, it is a goal, and he scores three points.

Kinvara had had a bad year with some tragic deaths, and the villagers really needed their spirits lifted. I appealed to the fairies to let Kinvara win. In the last minute of overtime, Kinvara scored a goal and won the match. Later on, I asked Kealan, a Kinvara team member who happened to be a staff member at Burren Nature Sanctuary, "Who scored the winning goal?"

Nobody knew…

Nature Magic

PART TWO
TOOLKITS

NATURE LOVER INTERVIEW: CATHERINE FARRELL

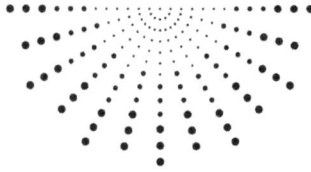

It's about putting the resources in the right places.

— CATHERINE FARRELL IS A RESEARCH FELLOW AT TRINITY COLLEGE IN DUBLIN
SPECIALISING IN RESTORATION ECOLOGY.

HOW DID YOU BECOME A NATURE LOVER?

I was hooked from an early age.

I guess I just love being outdoors. Both my parents had a good, strong respect for nature. I started to explore the esker that ran along the bottom of the farm. And I used to notice the orchids and the grasses and the water rushes; and then a beautiful hazel woodland with yew trees. I just loved being there.

WHAT IS YOUR FAVOURITE PLANT OR ANIMAL?

This morning, I went for my walk on the Grand Canal. The swallows were swooping down on the water. It's often the most common local things that I adore.

If I was to pick something out, it would be wetland birds, that collective of curlews, skylarks, lapwings... The curlew cry is so evocative. When you hear them in the summer, you know you are near a healthy habitat.

DO YOU FEEL SPIRITUALLY CONNECTED TO NATURE?

> I have to get out into the woods and immerse myself into it, just to re-set myself. There is that sense of peace... This sense of being part of something bigger than ourselves and feeling part of it; not feeling disconnected.

> That reminds you why I'm even trying to account for nature.

WHAT CAN PEOPLE DO TO HELP NATURE?

> Just get out and take five minutes and see what bees are buzzing around and what butterflies there are. Or, you can't beat being with someone who knows what they're talking about. Go on a guided tour.

IF YOU HAD A MAGIC WAND, WHAT IS THE ONE THING YOU WOULD YOU DO TO HELP THE PLANET?

> I think I would get rid of plastic and pesticides. I would replace them with naturally created materials. I think that would make a huge difference.

> I think that this pandemic that we're in can be the catalyst or the magic wand for change. Let people work at home a couple of days a week; don't throw them into long traffic streams... Let them be at home with their families. Let them hear the birds' songs in the morning when they're having their tea outside in the garden.

Visit www.naturemagic.ie (or search 'Nature Magic' on all major podcast platforms to listen to the full interview).

Visit www.incaseproject.com to learn about the INCASE project.

Check out Catherine's novel, The Easter Snow: www.catherinewilkie.ie

TOOLS FOR RUNNING A NATURE SANCTUARY

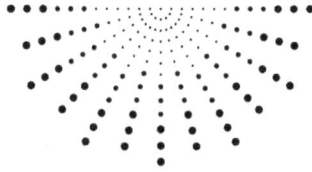

INTRODUCTION

Over the last seven years, since Burren Nature Sanctuary opened, we have run many walks, activities and workshops to engage people with nature. We believe that all this work could easily be replicated, with adjustments, anywhere in the world.

To help people who may struggle to design, perhaps a school tour or a group walk, we are sharing our outlines and any other information we feel you may find useful in this section.

The toolkits are just suggestions from what we have found to be successful. Feel free to replicate any of these activities and give us some feedback as to how it went! There are also some tips about opening and running a nature centre. For more details on this subject, please contact me through the website at https://www.naturemagic.ie/contact

INTERPRETIVE SIGNAGE

Good interpretive signage is key to educating people. The images must be excellent quality and professionally laid out, with text of the correct size for the sign.

Find a sign printer that has examples of interpretive signage or use a graphic designer who can give you a sample of their work.

In 2018, we won The Burren and Cliffs of Moher Geopark award for Interpretation. We get many compliments from visiting designers about signage and interpretation, so we hope the examples in our bonus material will help anyone who embarks on this task.

Stone Engraved Signs

These work very well for directional signs. The letters are engraved and can be painted. They are a lifetime job and always look natural and fresh. We have directional signs around the nature walk: Fairies, Round Field, Gypsy Wagon, Turlough, Home, etc.

We also have five stone 'Habitat Hunt' signs: Hazel Woodland, Ash Woodland, Shattered Limestone Pavement, Orchid Rich Grassland and Turlough. Each sign has the outline of a plant engraved into it and is painted white. Children can bring a crayon and take a rubbing of each sign and run on to find the next one until they have collected them all.

We have five stone signs set into the grass in the playground as stepping stones. They have the names of the Burren mountains on them: Ought-mama, Mullaghmore, Slieve Carron, Abbey Hill and Turlough Hill. To get around the assault course without touching the ground, you have to use the stones. Hopefully, those magical mountain names will lodge in the minds of young people and they will go in search of them one day.

Burren Flora and Fauna Aluminium Signs

- We have a set of twelve main interpretive signs illustrating all the main fauna and flora of the Burren. They include a section on farming in the Burren and water quality. A great deal of work went into the sourcing and formatting of the material for these signs.
- We have an Introduction to the Burren sign in the Waterfall Garden that explains the geological origins of the region.
- We have a map of the globe in the Burren Botany Bubble showing where all the plants in the Burren originated from. This sign includes the smells of the Burren! It is easy to set up compartments behind your signs for people to smell. We have drilled four holes and put a box behind our interpretive sign in the Burren Bubble. It contains four sugar shakers—the metal shaker lid with holes—just flush to the front of the sign. These are 'the smells of the Burren.' We add wild thyme, oregano, wild garlic and rose petals. During the winter, if the smells fade away, we top up it up with a drop of essential oil on a kitchen sponge.

- We have four seasonal flower signs around the nature walk: Flowers you might find in spring, summer, autumn and winter.
- We have a sign at the top of the ancient boreen explaining the history of the farm with historical maps.
- We have a sign at the Famine Village explaining the history of the ancient dwellings, a sign at the Round Field explaining the ancient farming methods of our ancestors, and a sign at the Turlough explaining the hydraulics of the Disappearing Lake.
- See all our interpretive signage in the bonus material: https://www.naturemagic.ie/bonusmaterial

AUDIO-VISUAL PRESENTATION

With grant aid from the Geological Society of Ireland, we made an audio-visual presentation for our visitors. The presentation explains the biodiversity at the Burren Nature Sanctuary and links it to the rest of the Burren hills.

It is a beautiful 15-minute piece and a great tool for visitors not booked in for a guided walk. Everyone has 15 minutes to sit down, enjoy and learn. Our wish would be to have a dedicated area to play the film. At the moment, it is on a flat screen TV that we move around to whichever room is vacant (which is not ideal).

The Script of Our Short Film

Intro

The Burren is a rocky place. Beneath these rocks is a mysterious world of caves and disappearing lakes. Above these rocks is a botanical wonderland, where wildflowers and orchids bloom, where plants and trees grow close to the ground, spreading out to make the most of the thin soil.

The story of the Burren landscape begins far back in geological time, to ancient tropical seas, to the earliest human settlers who hunted and gathered and then farmed this land; all the way up to those who live, work, visit and care for this region of spectacular natural beauty. With a remarkable landscape, diverse and unusual flora and fauna, the Burren is a place where people come for inspiration, tranquillity and restoration.

The Burren is a region along the West Coast of Ireland, spanning parts of counties Clare and Galway. The vast slabs of limestone pavement, up to 800 metres thick in places, stand as monuments to the ancient tropical seas in which they were formed 350 million years ago. The Burren was once the bed of a warm, tropical sea. The remains of ancient sea creatures deposited on the seabed formed into the calcareous limestone rock that forms the stark beauty of the Burren. Look closely to see fossilised corals and snails on the rocks.

Calcium-rich limestone is soluble by the mild acid in rainwater. Rainwater erosion of limestone has worn channels in the rock and small pools where water and wind-blown soil gathers. This process results in the characteristic appearance of the Burren: huge crevices in rocks, pockmarked and shattered. The smooth surfaces are termed clints, while the cracks that separate them are called grikes.

Plants, Flowers and Trees

At first glance, the Burren may appear to be a barren wasteland, a place of stone, devoid of life. But look a little deeper: between the rocks, plants eek out an existence, clinging on to life by placing their roots into the deposits of wind-blown soil that settle in the cracks in the rocks; the iconic spring gentian with its blue colour. The overwhelming ability of nature to reclaim is evident in the Burren and it is no wonder it has been named 'the fertile rock'. If these rocks were left ungrazed, scrub would encroach and the rocks would be covered with bushes and then trees, dominating the landscape and preventing the unique Burren flora from growing.

Turlough

Another geological wonder of the Burren region is the disappearing lakes. Turloughs or disappearing lakes come from the Irish 'tur lough', meaning 'dry lake'. The turlough at the sanctuary is extremely rare. Although it is fresh water, it is also tidal. In summer, the lake drains and fills every 12 hours (in time with the tides in nearby Kinvara Harbour/Bay). When the turlough is dry in summer, plants and flowers such as the creamy white meadowsweet, purple loosestrife, water mint and forget-me-not carpets the turlough. In winter, there is always water present but the levels fluctuate dramatically and it is home to visiting, migratory birds such as widgeon and Eurasian teal.

Woodland

At one time, Ireland was completely covered in woodland. Native Irish trees include ash, hazel and oak. Throughout history, woodlands have been cleared for agriculture, industry and house building. In pockets of the Burren, ancient stands of ash and hazel remain. It is evident that humans have been farming this landscape for thousands of years. The stone monuments such as Poulnabrone Portal, the ring forts and stone forts and the stone walls were in continual use. Ash trees grow well in limestone soils, it is recognisable by its pale timber and black buds. The wood is used to make furniture and, notably, to make hurley sticks for the national sport, Hurling.

Today, hedgerows are important. Often called linear woodlands or green highways, the boreen at Burren Nature Sanctuary is bordered by whitethorn, blackthorn, spindle, ivy and elder. In early spring, primroses and other flowering plants bloom. These native Irish trees provide essential food and shelter for wildlife: small birds and mammals such as robins and hedgehogs. This boreen has been in continual use until the last century and was used daily to bring cattle and sheep to the fresh water in the turlough.

Meadow

The meadows of the Burren are a botanical haven. Having been cleared by our first farmers during the Neolithic era, low level grazing by animals meant that no one grass became dominant and therefore, in spring and summer, the array of wildflowers and orchids can bloom. This relationship between traditional farming practices and the flowers of the Burren is central to the Burren story. Farming continued unchanged for thousands of years. The rocky land is not suitable for tillage, so livestock farming was the main. Hay is cut in autumn, when flowers have seeded, to maintain the flora in the meadow. Over winter, animals graze the grass, allowing the wildflowers to compete with the hardy grasses. The practice of transhumance or winterage, where animals are brought up to the hills in winter, is central to the success of the wildflower meadows. Seeds and bulbs lie dormant in winter, undisturbed by chemicals. And suddenly in spring and summer, a riot of colour emerges: daisies, buttercups, clover and many other wildflowers. In high summer, the meadow is rich with orchids that thrive in the limestone rich soil: early purple orchids, pyramidal orchids and O'Kelly's orchids.

Although Ireland did not have a true industrial revolution to the same extent as other parts of the world throughout the 18th, 19th and 20th centuries, there has been a definite move away from the land toward larger towns and cities and a decline in farming. The Burren is truly a cultural landscape, formed through geological forces, influenced by humans who removed trees, grazed animals, built stone walls and giant monuments to their ancestors.

Today in the Burren, farmers who recognise the unique biodiversity are continuing the traditional farming practices of winterage. These activities maintain the balance of nature between rocks, flowers, grasses, animals and the communities who live, work and visit here. The Burren is truly a cultural landscape.

THE AUDIO TOUR

We created an audio tour app for the Burren Nature Sanctuary. There are ten stops that explain the biodiversity and habitat of each one. A soothing, lyrical voice presents the information.

The pros of an app are that you need not hand out or rent an audio device for people that want to have a tour when there is no guide available. People can download the free App in the café with the free Wi-Fi.

The drawback is that people are then staring at their phone as they walk around. This is not ideal as you want them to disconnect from technology.

We made the app ourselves using Mobincube and pay a small monthly fee to run it on their platform. It is simple, and it works. Don't pay an app developer who will charge you thousands! Just record your tour in 30-second sound bites and load it up!

You can do this for your local park, village or beauty spot to point out local plants and birds. Check out the app here:

https://apps.apple.com/ie/app/burren-nature-sanctuary/id1104975765

Introduction

The introduction is a welcome message explaining that the Burren is a UNESCO World Heritage Site. It elaborates on the size of the Burren and talks about its diversity. It then explains that Burren Nature Sanctuary is an interpretive centre of the Burren, designed to inspire visitors to cherish and protect the natural environment. We encourage people to enjoy our one-mile nature walk and to enjoy fresh, local produce in our award-winning café. It also mentions the Waterfall Garden; we draw special attention to the flora growing there and the fossils in the limestone pavement.

After the introduction, the audio tour takes our visitors through the following ten stops and explains the significance of each:

The Burren Botany Bubble housing the National Living Collection of Burren Flora.

The Boreen is a grassy track. It was a busy thoroughfare in use up to only a hundred years ago; it was used to bring cattle and sheep to water from neighbouring farms.

The Disappearing Lake is our extremely rare freshwater tidal lake that drains every 11 - 12 hours.

The Meadow rich in seasonal orchids and other plants which thrive in the limestone rich soil.

The Doline is a collapsed underground cave. Apparently, underground souterrains led from this farm all the way to Dunguaire Castle in Kinvara.

The Ash Woodland gives visitors a glimpse of what the Burren would look like if left it ungrazed by cattle.

Shattered Limestone Pavement: our sanctuary has 25 acres of undisturbed karst limestone shattered pavement—home to an abundance of Burren wildlife.

Hazel Woodland/The Fairies: explore the fairy houses in the hazel woodland opposite the ancient dwellings.

The Ancient Dwellings is a farm complex with evidence that humans occupied the area during the famine period and possibly as far back as the Iron age.

The Round Field was historically used as a holding pen for cattle and sheep.

The Farm Pets: We encourage visitors to visit our friendly farm animals.

SUNDAY SERIES WALKS

In summer we mostly take botany walks. An experienced guide is essential.

For winter, we designed three walks based on:

- History,
- Geography,
- Palaeontology and
- Glaciation of the area.

We took these walks in rotation, encouraging people to come back for three weeks in a row.

These examples of the guide notes show the depth of knowledge needed to take a group of adults on an interpretive walk:

http://www.naturemagic.ie/bonusmaterial

TOUR GROUP EXPERIENCES

It is a certain skill to present and sell experiences to tour group companies.

The information must be clear, concise and enticing. The tours must offer excellent value, with a sliding scale of discounts (depending on numbers booked or whether the company is 'reselling' the tour to another company).

Failte Ireland ran a brilliant course for tourist businesses to train them how to layout their 'saleable experiences'.

The pdf of our experiences is in our bonus material, but the top tips are:

- Images sell the product. Make sure the images are appealing and show exactly what you are offering (easier said than done).
- The introduction is a brief history of the attraction allowing an insight into the people.
- Highlights elements of the offering or site.
- Each experience offering has images, a short-tabulated description with prices and a longer explanation of the running order.
- At the end of the document, show general information including food offerings and prices, capacity, venues, opening hours, etc.

On our sales document, we list these experiences:

- General Experience.
- Express Guided Tour.
- Premium Experience.
- Family Experience.
- Scone Making Demonstration.
- Wild Rose Skincare Workshop.
- Fairy Pig Walk.

Check out our 'saleable experiences' document:

http://www.naturemagic.ie/bonusmaterial

FOCUS ON BOTANY: BE A CITIZEN SCIENTIST

This is an example of a short guided welcome and introduction to Burren Nature Sanctuary and botany that takes 15 minutes in the Burren Botany Bubble.

A member of staff briefly explains the botany of the Burren and shows people how to record plants for the National Biodiversity Centre Database.

The member of staff gives a handout with instructions and then encourages visitors to set off around the walk to record plants or to do this on walks when they go home.

Here is more information on the Biodiversity Database website and a user-friendly app:

https://apps.apple.com/ie/app/biodiversity-data-capture/id906361120

This is the outline of the talk:

The Burren is a special place for botany. The rare and unusual wild orchids draw botanists from around the world.

How to help Ireland record its natural heritage:

- take a photo of the whole plant,
- take a photo of the habitat,
- take a close-up of the flower,
- take a photo to illustrate scale (ideally a ruler but you can improvise with, for example, a coin or your hand) and
- tweet/email the photos and your full name to @burrensanctuary or hello@burrennaturesanctuary.ie

If you cannot identify the plant, we will identify it and log the record with the National Biodiversity Database. (http://www.biodiversityireland.ie)

(For help with identification, see http://www.botanicalkeys.co.uk)

Photos can be sent straight to the database, but we like logging what people find on-site, so we encourage them to share their pictures with us.

NATURE LOVER INTERVIEW: EAMON RYAN

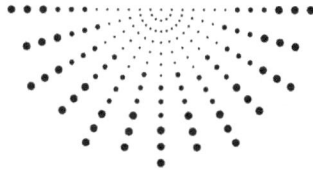

Every place matters, every person matters.

— EAMON RYAN IS T.D. (TEACHTA DÁLA, AN ELECTED MEMBER OF DÁIL ÉIREANN—THE IRISH GOVERNMENT) AND THE CURRENT LEADER OF THE Green Party IN IRELAND.

HOW DID YOU BECOME A NATURE LOVER?

I wasn't a great student, but I was lucky. At age 16, instead of doing the Inter-cert, we did a series of O levels and one of those was Ecology. This was 1979, and we were reading Gaia Theory, Silent Spring and Limits to Growth and all the latest ecological thinking at the time. It really got to me and I had a sense of the interconnectedness and the urgency around the need to change.

WHAT IS YOUR FAVOURITE PLANT OR ANIMAL?

I don't think it's a plant or an animal, it's more a place. Inishbofin, on the West Coast, out in the Atlantic. I love it. I do a lot of fishing, swimming and snorkelling; and it's that connection to the sea that really inspires me.

WHAT CAN PEOPLE DO TO HELP NATURE?

A farmer in Clare who was involved in protecting Mullaghmore, Patrick McCormack, said something simple, that I thought was very good:

"Walk, get out and walk, walk the land." That sounds very simple, but it is my first recommendation.

Also little steps, like planting stuff. It's good for physical health, mental health and that sense of connection. I think it's those simple, local, practical things without it being a big 'virtue'. Small things like that are a start. There is something healthy in getting out in nature; it's really good for you.

IF YOU HAD A MAGIC WAND, WHAT IS THE ONE THING YOU WOULD YOU DO TO HELP THE PLANET?

One thing I would I do today is a national land use plan. That sounds very prosaic, but I would like to address what we can do in Ireland, where we have some control. It would recognise that human beings are part of nature, we are not separate. We would be planning a rural Ireland that is vibrant, sustainable and secure; a great place to raise families and to be healthy. But you have to look at the layers of the plan—restoring biodiversity and reversing the losses from the last 50 years.

Every place matters. Every person matters. Maybe it's that understanding which will change our ways.

Visit www.naturemagic.ie (or search 'Nature Magic' on all major podcast platforms to listen to the full interview).

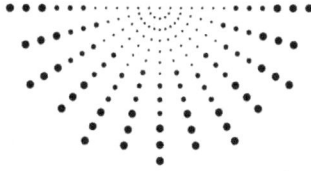

BASKET MAKING

Introduction

Basket making is my top choice for a fully engaging adult activity. We have run basket making courses and they have all been fulfilling and healing.

This is my favourite activity of all time! Everyone enjoys it.

Equipment Needed

- Willows.
- Tools.
- Individual tables, with at least 1 m between each.
- Protective glasses.
- A tub for soaking.

Method

Willows need to be conditioned correctly for basket making. This means soaking in a tub for several days.

Lessons We've Learned from Basket Making

- We reconnect with our ancestors (who all used baskets). Working with the natural material is harmonious; creating the basket involves complete mental and physical focus. In this way it is an ideal meditation that brings you instantly into the present moment. The activity takes you away from your day-to-day life as it needs all your concentration. You leave with a beautiful, natural artefact and a sense of achievement having learnt a new skill.

- In one stage of making the baskets, sharp sticks are pointing in all directions! You need at least a metre between tables.
- To hire a basket maker with equipment and all supplies is expensive.
- As it is an expensive class to run, we recommend pre-booking with the class fee paid in advance (and a refund policy).
- If you become proficient in basket making, it would substantially reduce the cost of running the activity. However, you would need all the equipment and a supply of willows (which we have planted!). I have done 3 courses and still struggle to remember the process without Ciaran Hogan in the room!

WILD ROSE SKINCARE WORKSHOP

Introduction

This is a lovely experience and a good option for rainy days. It connects people to nature and engages them by letting them create their own beautiful product, which they can take home as a memento or a gift.

Even on the worst day, people get to pop out to the Botany Bubble and see the Native Herb beds and take photos of the Herb Signs.

Visit https://www.airbnb.com/experiences/1565656 to check out our Airbnb experience.

If you want to run a similar activity, here are some recipes to get you started:

Wild Rose Bath Salts

Ingredients:

- 200 g coarse salt (we use local Atlantic salt).
- 5 g dried rose petals (optional-chamomile, carrageen seaweed).
- 10-20 drops essential oil.

Method:

- Measure your salt.
- Pour into the mixing bowl.
- Smell the essential oils and decide which fragrance you like.
- Add a few petals if you like or some seaweed for the natural oils.
- Pour into your jar and label!

To Use:

IMPORTANT: Do not use if you have high blood pressure.

- Add a handful of salts (100 g) to a full body bath.
- For a foot bath, add 50 g.
- Add 5 - 10 drops of essential oil.
- Suggested containers: reuse jam jars.
- Shelf life: 18 months (if moisture free).

Rosehip Face Serum

Face serums are a very popular gift. They are expensive to buy but easy to make with high quality oils. Sunflower oil is an inexpensive, widely available ingredient (buy organic) and soaks into the skin without leaving an oily residue.

Rosehip oil is suitable for all skin types, has anti-aging properties, helps to reduce scarring and is useful for acne, burns and mature skin.

Ingredients:

- 14 ml organic sunflower oil.
- 100 drops of Rosehip oil.
- 2 drops Rose Geranium essential oil.

Method:

- Add 100 drops of Rosehip oil to your dropper bottle.
- Fill up to below the neck with sunflower oil.
- Add the essential oil.

- Close the bottle tight and roll between your hands to mix.
- Clean the bottle with lemon/vinegar spray.
- Write your label and attach it.
- Suggested containers: 15 ml brown glass dropper bottles.
- Shelf life: 6 months (if moisture free).

Whipped Shea Body Butter

Note: this is easy to make, but it requires equipment.

I suggest, to reduce costs, you demonstrate how to make it and not allow your guests to join you.

Ingredients:

- 100 g shea butter.
- 100 g of rose petal infused organic sunflower oil (or face serum for face cream).

Method:

- Put the shea butter and sunflower oil into a food processor. Make sure it is clean and dry.
- Blend on lowest setting until smooth.
- Remove from the food processor and put in a bowl.
- Whisk for a few minutes with an electric whisk. This adds air and makes it fluffy.
- Spoon into jars (tap the jar to settle the cream so you can fill the jars).
- Wipe off with kitchen towel, then wash it in hot water with washing up liquid.
- Dry.
- Write label and attach it.
- Suggested container: 60 ml brown glass jar with a lid.
- Shelf life: 6 months.

Note: bacteria will not grow without water, so be careful that all your equipment is dry.

How to Make Infused Oil

- Cover the dried plant (in this case rose petals) with a base oil (in this case organic sunflower oil) in a (dry) glass/metal bowl.
- Place over a simmering pan of water to warm the oil for 4 - 12 hours and slowly extract the properties of the plant.
- Drain the oil through a muslin cloth and squeeze well!

How to Dry Plants

- Pick plants and leave in an air dryer in a cool room, out of direct sunlight.

THE MAGIC OF TREES

Introduction

This is one of our Airbnb experiences. It helps people clear their heads and reconnect with nature.

Learn more about this experience here:

https://www.airbnb.co.za/experiences/579926

The activity also introduces people to the ancient Ogham language. The ancient Irish people used a language with symbols called 'oghams'. It consists of upright lines with various combinations of lines intersecting at angles.

When children see the symbols, they instinctively sign out the letters with their fingers across their arms. It may have originated with a sign language using fingers placed across the arm to indicate a particular letter. It is found scratched on tablets and standing stones across Ireland, Scotland, Wales, Cornwall and the Isle of Man. Each Ogham corresponds to a tree or plant. The ancient people had a 13-month calendar, each one represented by a tree.

Forest Garden

The wonderful Mary Reynolds calls herself a 'reformed landscape designer'. She wants everyone to give half of their gardens back to nature. (See her interview in the Traditional Herbal Remedies chapter.)

She has written the amazing book: The Garden Awakening. It details how to design an edible forest garden, using layers of trees and ground covering plants that need no maintenance and are a yearly source of food.

During the 2020 lockdown period, we designed and started a forest garden. So far, we have planted three oak trees and 15 hazel trees in the north-eastern corner of the chosen plot. It is a project that we can build on year by year; and this autumn we plan to plant the east boundary with a mixed hedge of berries. The garden will then be completely sheltered.

When we have the resources, we hope to add the paths, etc. The joy of this type of gardening is that it works with nature and you can start a forest garden slowly and built on it yearly.

We aim to add the following to our forest garden:

- A mulberry tree.
- A sweet maple tree.
- A sweet chestnut tree.
- 3 plum trees.
- 3 honeysuckle trees.
- 6 pink hawthorn trees.
- 6 barbaris bushes.
- 6 raspberry canes.
- 6 tayberry canes.
- A meditation seat.
- Paths.
- Sun-trap seat.
- A pond.
- Some guinea fowls.

Inspired by the book The Garden Awakens by Mary Reynolds.

SCHOOL TOURS

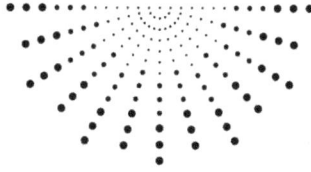

LEAVE NO TRACE SCHOOL TOUR

Introduction

School tours are roughly three hours long. We have found that most school tours get booked from 10 AM to 1 PM.

On the very first school tour that we ran, the national radio arrived to interview me as I lead the tour. Waiting for a bus of 40 children to teach them Leave No Trace principles and leading them around a mile-long nature walk with a microphone stuck in your face, is intimidating (to say the least)! However, it went brilliantly and the school tours have been an enormous success ever since.

Over 15,000 school children have completed the Leave No Trace training at Burren Nature Sanctuary.

Leave No Trace is a code of outdoor ethics with seven principles (see end of chapter). Each principle has an accompanying action that each child engages in.

The children wear bibs with the principles on the front. As they perform the actions, they take a photo for discussion in class later. For example, the principle 'Respect Animals and Wildlife' may be a child holding a guinea pig. There are three sections to the tour: nature walk, lunch and play.

These are the steps to running a successful school tour:

- The school books the tour. We do this through an online calendar for a very important reason. It ensures that the school secretary enters the correct day, the correct number of children, and if there is a mix-up, the responsibility is with the school and cannot be blamed on a member of our staff! There is nothing worse than finding you have double booked 50 kids or they show up unexpectedly! We don't take a deposit on booking as children normally bring the money to school on the day and the school would have to send a cheque from their own funds which would confuse their bookkeeping systems. We have very few cancellations. Once they have booked their bus, they are coming!

- After booking, the school receives a pre-visit pack. This is a lesson plan for the teacher (teachers love a lesson plan!). The lesson plan explains Leave No Trace and includes a slideshow for a discussion of the principles.
- The bus arrives. There is a tight format for the day to ensure it runs smoothly. This is explained and discussed with the teachers. We can adjust it slightly if they want to work around snack times or bus times. The teachers fill out the names of all the children and sign a sheet with some instructions about play equipment in the playground. This form becomes essential when the head teacher comes to pay as it removes the debate as to how many children are on the tour! We try to do the nature walk first. That way, everyone has burnt off a bit of energy, is hungry for lunch and can relax and play for the rest of the day.
- The first principle is 'Plan and Prepare' and the lesson is 'Go before You Go', so anyone who needs the toilet goes immediately!
- We hand out the bibs one by one asking, "Who knows what this principle means?" If someone answers, they put it on. Older kids may not be keen on wearing the bibs as they don't want to look uncool, so we have the principles on a set of cards in the rucksack in reserve if that happens. We have a pack of equipment for the walk that includes a tablet (with a protective, waterproof case) for taking photos, bug viewers, binoculars, a butterfly net and butterfly and bee identification books, etc. We hand out the equipment and tell the children they will all get a turn. Sometimes we take Frisky, the goat, on a leash.
- We stop half way around the walk at the Gypsy Wagon and have drinks ready for the children (this is important on hot days). We

ask them to imagine they are living a hundred years ago and there are no supermarkets. How would they live off the land? How would they live in a wagon?

- On the walk, we stop at various points, and discuss the Leave No Trace principles and look at bugs and birds that the children may have spotted. Children carefully collect bugs/bees/butterflies in the bug collectors and nets, and we try to identify them with the reference books we carry in the rucksack. They learn about the history of the Burren and why the biodiversity is so great in this area. They learn where they are in relation to the County, the Country and the world (how many school tours did you go on that you hadn't had a clue where you went?). They visit the Fairy Woodland and get to feed the animals. We try to teach them the names of a few plants. The way to do this is to ask them if they know the name, get them to repeat it and then spot it again and ask them the name and get everyone to repeat it again. If they say it three times, they remember for life.
- Then it is lunchtime. Children will have a packed lunch. They sit down to eat (important point!).

When designing school tours, we keep the following Confucius quote in mind:

> I hear and I forget. I see and I remember. I do and I understand.

Challenge: Teachers look for free tea or coffee and cakes, and sometimes, a full free lunch. We now offer a cup of tea or coffee per teacher and a scone. If there are extra adults on the tour, we do not provide free catering to them. It all adds to the cost and margins are tight. Make sure this is clear on the pre-visit information.

Challenge: Most teachers are brilliant, but some just want to enjoy a day off while someone else looks after their class. This puts more pressure on the staff as the group requires more supervision and the children are normally less happy and look for attention by acting out.

After lunch, children get to play in the adventure playground and in the indoor soft play area. We aim for everyone to get time in each area. Staff monitor the play areas.

At the end of the tour, we give the teachers a feedback form with some questions:

- How would you rate the tour out of 10?
- What was the best bit of the tour?
- What was the worst bit of the tour?
- Were you happy with the pre-lesson plan?
- How happy were you with the staff? Any comments welcome.
- How would you rate the educational content out of 10?
- Would you return with a different class for an age-tailored Leave No Trace school tour?
- Any other feedback welcome, thank you!

This means we can deal with any issues immediately.

Each child then gets a Leave No Trace certificate and a Welcome Back voucher. This is 50% off for a family day pass and we usually see the families later in the summer.

After the tour, we send the teacher a post-visit lesson plan with the photos (deleting the terrible ones!) via We Transfer. The children then refresh the principles in class as they look at their classmates engaging in them. (For more information on the principles, see https://www. leavenotraceireland.org) Become a member of Leave no Trace and design your own tour around the principles using our method!

Leave No Trace Principles:

1. Plan ahead and Prepare.
2. Be Considerate of Others.
3. Respect Farm Animals and Wildlife.
4. Travel and Camp on Durable Ground.
5. Leave What You Find.
6. Dispose of Waste Properly.
7. Minimise the Effects of Fire.

See the 'Leave No Trace' lesson plans in the bonus material:

http://www.naturemagic.ie/bonusmaterial

PEOPLE AND PLACE SCHOOL TOUR (AGES 10+)

A pre- and post-lesson plan are sent to teachers as with the Leave No Trace Tour.

On this tour, children learn curriculum topics of history, geography and ecology by engaging in role play at different points around the nature walk. Children also get to meet the animals and play!

The tour teaches children about:

- The early Stone Age.
- The middle Stone Age.
- Medieval Kinvara (950 to 1400 BCE).
- The periods after the famine (1740 – 1741 and 1850 – 1980).

Activities include roleplaying where children have to talk about what they would learn if they lived in the Stone Age, and the skills they would need to survive in the Stone Age.

See 'People and Place' here: https://www.naturemagic.ie/bonusmaterial

GEOLOGICAL FIELD INVESTIGATION (AGES 12 - 15)

This school tour was designed to meet the Geography project criteria for the Junior Cert (children age 12-15). See 'Geological Field Investigation' here:

https://www.naturemagic.ie/bonusmaterial

CYCLES OF NATURE (AGES 15+)

Cycles of nature was a three-part course designed for Transition Year students (ages 15 - 16) on three separate days throughout the year in different seasons funded by Galway County Council Heritage Department. Each day, we included learning a traditional skill and an animal interaction.

SESSION 1 (LEADER NOTES)

Introduction

During the introduction, we welcome the students to Burren Nature Sanctuary; give an outline of the Leave No Trace principles, explain our facilities, etc.

We continue to explain the purpose of the course: to show how the unique Burren landscape changes throughout the seasons of the year and how the changes would have influenced people living here in the past.

The introduction is followed by an explanation of how we will observe the area with its different habitats over the following three seasons.

We will do this by:

- Recording the changing environment at set points on the route during each visit over the following seasons. We will keep photographic records, journals, and collect environmental data. We will compile the collected data and use it to monitor long-term changes of both the biodiversity and the environment of the area.
- Conducting a grid square survey in the meadow. This will give us a chance to see and record the variety and timing of the amazing range of both grasses and flowers that grow in this traditionally maintained grassland.
- Learning both new and traditional skills. We will learn compass and mapping skills to map the changes in the area. We will learn to identify and catalogue the amazing flora and fauna we have here. And each session will give you the opportunity to learn a

nature craft associated with the season, using the natural materials around you.

- You will submit a project at the end of all the sessions. This can come in any media—soft copy or hard copy. But we will not accept presentations devolved through the media of interpretive dance or mime! There will be an award for the best project!

Today:

- We will get togged out in all the weather suitable gear you have brought and walk the route.
- You can become acquainted with the area and see the selected locales for taking your photo records.
- We will do a grid square analysis in the meadow.
- We will get back here and you will have time to try your hand at some pottery. On your walk, you can collect some natural materials such as bark, a leaf or some grasses to incorporate into your pottery.
- We will end with tea and biscuits.

Animal Interaction: Pig Paddling (try to steer the pig around a course of cones; see page 18).

Photo Locales:

1. Fairy Woodland: Fairy Post Box facing the red fairy door.
2. Pinch Point in path near fairy woodland exit. Facing the Fairy King's door, two trees in frame. (Will get them to take a compass bearing at this point.)
3. April & May Flora Sign 330 degrees NW.
4. Viewing point at shattered pavement facing 90 degrees, due E.
5. Oregano bend facing 120 degrees SE
6. July Flora sign facing 360 degrees, due N.
7. Doline facing 30 degrees NE.
8. Orchid meadow at the Burren fauna sign facing 90 degrees E and 180 degrees S.
9. Grid square exercise in the meadow.

10. Double gate facing 340 degrees N and 15 degrees SE
11. Wooden gate facing 100 degrees E.
12. Turlough sign facing 38 degrees NE and 360 degrees N.

At each point, discuss how they think this environment will change, and why the locale was selected.

SESSION 2

Revisit the recording sites from session 1.

Nature Craft: Make Pesto

We will forage Ransoms (wild garlic) and use them to prepare a wild garlic pesto. We will use the crushed leaves, oil and cashew nuts to make the pesto and will eat this with bread. While not a traditional dish, it is a good way of using the fresh local plants. We will learn about traditional herbal remedies.

Animal Interaction: Put on Cookie's (the donkey) traditional willow baskets and bring your lunch to the Gypsy Wagon, leading Cookie without scraping the baskets against the bushes.

SESSION 3

Revisit the recording sites from sessions 1 and 2.

Candle Making

This session will give you the opportunity to learn a traditional skill associated with the season. As the bees have become more active, we will use beeswax to make a cream and balm. We will also use beeswax to make candles.

Animal Interaction: Take Frisky (the goat) for a nature walk.

SCHOOL TOUR ACTIVITIES

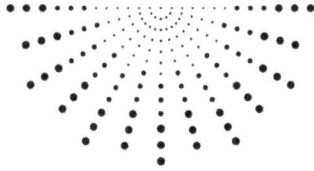

GYPSY WAGON IMAGINATION

On our school tours, we take a break in our Gypsy Wagon Shelter.

After a refreshing drink, I get all the children to sit down and play "imagine". (Never speak 'at' children.)

Ask them questions, how much they know will surprise you. This exercise can really connect children to the land. They have an innate ability to adapt and survive and yet they are so stuck in their reality…

We imagine we are Traditional Travellers 100 years ago, moving from village to village in our wagon with our horses. There are no shops around.

- How do we get our food? (Common answer: Catch a cow! Talk about hedgerow food—blackberries, etc.)
- How do we cook our food?
- Where do we get our fuel?
- How do we earn money?
- What do we feed our horses?
- Where do we get our horses?
- Where do we get our clothes?
- What about a doctor?
- Communication?
- No plastic bags. How do we carry things?

THE FABULOUS FIVE

This is a fun activity that can be done outside (or inside on a rainy day). Kids love the hunt and the fact that they get to keep the bracelet as a memento.

Top Tips:

- Get the kids to line up to collect a bead so you don't get trampled!
- Hand the beads out—it's a tedious task picking a spilt bag off the floor!

Children follow the clues and collect 5 different coloured beads and thread them onto a pipe cleaner to make a bracelet. The five beads represent the things plants need to grow: sun, water, air, space and soil.

What you need:

- A bag of yellow beads, blue beads, white beads, green beads and brown beads.
- Five clues.
- Pipe cleaners.

Clues:

- Image of a sun.
- A jar of water.
- A pinwheel or image of a windmill.
- Image of plants spaced in a row.
- A bucket of soil. (This can be the start of the 'plant a seed' activity coming up next.)

Show the children the seed.

Clue 1: Sun

In order for me to grow big and strong

You are going to need to help me along

Five things I need to stay alive

We'll call them the fabulous five

The first will surely help me wake

It's cold in here for goodness' sake

I must warm up and feel the light-

Take me where it's warm and bright

Clue 2: Water

I'm much warmer now, thanks a bunch,

But I think it's getting time to munch.

I make my own food whenever I'm hungry.

But the problem is, I'm really thirsty!

Look around - you need to think,

And find something for plants to drink

Clue 3: Air

You need me, and I need you!

Soon you'll learn a step or two

We eat and drink and need to share,

'Cause both of us must breathe the

_____

Look around - think and observe.

Can you see where wind is pushing air?

Clue 4: Space

Even though I'm little now

I'll soon enough be big - somehow.

I'll grow with others (it's not a race)

Just don't plant me too close,

I need my_____

Look around! You'll see a sign,

Showing plants with room, growing fine.

Clue 5: Soil

Sun, water, air, and space-

Are things I need to grow.

But there's one more thing I need, you know.

It's dark and brown, under your feet.

Without it, my life will be incomplete!

Look around — a bucket and a sign!

Complete the "Fab Five" and your plants will grow FINE!

The Fabulous 5:

- SUN
- WATER
- AIR
- SPACE
- SOIL

Now plant your seeds!

Thank you to Linda and Carol Wellings and the Shelburne Farms team for allowing us to share this great activity. From their wonderful book, Cultivating Joy and Wonder. www.shelburnefarms.org

MAKE A PLANT POT AND PLANT A SEED

This activity is very popular. The challenge is… It is messy! You need a box to give to the teachers to take all the pots home.

This is an ideal activity to do indoors (in a barn if possible) if the weather is bad and follows neatly after The Fabulous Five. To extend the activity, use a large drawing of the parts of the plant and get them to pin on the names of the different parts of the plant.

If there is a tablet/laptop, show them a time-lapse of a seed sprouting.

To Run the Activity:

Demonstrate making the pot

Children are pretty good at waiting in line. Divide the class up into however many plant pot formers you have (we have five).

Form the pot.

Move on to the next table to fill the compost (a teacher can be in charge of the bag!).

Move on to the next table to plant the seed.

At the end of the activity, explain that the plant will need to be watered and when it is 6 inches high, they can plant it in the garden.

Suitable seeds:

- Sunflower.
- Marigold.

WEB OF LIFE

This activity can be adjusted for your own local environment. It is a great activity for a large group of people (of any age).

All you need are the cards with animals on them and a ball of string.

- Hazel tree
- Quaking grass
- Bramble (Blackberry bush)
- Spring gentian
- Wild thyme
- Ash leaf
- Feral goat
- Pine marten
- Fox
- Badger
- Hare
- Otter
- Peregrine falcon
- Cuckoo
- Heron
- Robin
- Snail
- Common carder bee
- Ant
- Spider
- Dragonfly
- Fish
- Human

Hand out the cards. Ask the person with the ball of string to throw it to someone they are connected to and to tell the group why (for example, feral goat to ash leaf—goats eat the leaves).

Keep going until all the string is gone. Some people will get the ball of string more than once.

A tight web is formed.

When you are finished, ask one person to drop the string. The whole web will collapse. When two or three children drop the string, the ecosystem has completely collapsed…

This fosters a respect for the natural surroundings and is a visual representation of how nature is connected. When one part is removed, the entire ecosystem suffers.

PICK A FLOWER

This is a very easy activity that illustrates the principle, 'Leave what you find.'

It is a lesson children and adults at Burren Nature Sanctuary should learn before they start the nature walk.

Print out 30 to 40 native flowers and laminate them. Spread them out on the floor. Ask the children whether it would be so bad if they picked one flower? Ask one child to pick a flower.

"That doesn't seem to bad huh?"

Then ask the children to pick one flower each.

"See how quickly all the flowers are gone?"

Explain that if you pick a wildflower, its seeds cannot make the baby flowers for next year. Ask them to explain this to their adults when they go on walks. They are always keen to boss around adults!

RUBBISH EXPERIMENT

We made a grid of 9 compartments with a piece of strong wire mesh and a gap where different types of waste could be left.

The rubbish experiment is simple and effective but needs a spot where it can be left undisturbed over a few years. Add some different types of waste to the compartments: paper, glass, a tin, a plastic bottle, woolly hat, piece of wood, and see how long they take to decompose. The aim of this activity is to make people aware of how long it takes for waste to decompose. When they see it 'in action', they will make more conscious decisions when it comes to what they buy and how they dispose of household waste. See 'The Rubbish Experiment' sign in the bonus material:

https://www.naturemagic.ie/bonusmaterial

BIRD BOX TREASURE HUNT

There are twenty 'bird boxes' around the nature walk with clues inside them. We designed this treasure hunt with an accompanying 'pack' that you could buy. It contains a card with images of the correct answers. Next to each image was a box which could be filled with the number of the bird box from that question. At the bottom of the card was the key to collect your treasure. You could then claim an ice cream.

Not many people buy the pack, but everyone enjoys the questions around the walk and searching for the next box! We find the cards useful for groups of scouts and sometimes for older students as an icebreaker. The concept of the bird boxes trail has worked well and proves to be a great incentive for children to complete the mile-long walk.

Instructions:

Find each bird box and use the clues to complete the treasure trail card. Get the code to solve the puzzle. See the clues and the answer card in the bonus material: https://www.naturemagic.ie/bonusmaterial

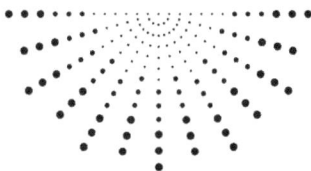

BEEHIVE DRESS-UP

This is a fun activity that really gets children talking. Most children have experienced wasp or a bee stings, or they have seen their parents running away from them, so they are naturally scared.

Explain why we like bees and all the different jobs they do. Tell the children to dress up as the different members of a hive, let them buzz around and look after the hive!

Encourage them to swap roles—so they all get a chance to be Queen! Discuss pollination and why bees are so important to humans.

Queen: The head of the hive, busy laying up to 15,000 eggs a day. (Paper crown)

Drones: Male honey bees who mate with the Queen once a year hang around, being fed by the female worker bees. (Several bow ties)

Nurses: Honey bees that make food, called brood food, for the young larvae or baby bees. (Several white paper nursing caps)

Guards: Honey bees who smell every insect that comes into the hive. If the pheromone is not the correct smell, the guard will chase the insect away. (2 or 3 cardboard cones to tie around the waist [stingers])

You will also need:

- Larvae (lots of stuffed old white socks).
- Whisk broom (to clean the hive).
- Egg cartons (to build honeycombs).

- Plastic jars (for honey making bees).
- Hand fans (to cool the hive).
- Trays (for the workers to bring food to the Queen).
- Baby bottles (to represent feeding the larvae).
- Small baskets (for the worker bees who leave in search of pollen and nectar).

Thank you to Linda and Carol Wellings and the Shelburne Farms team for allowing us to share this great activity from their wonderful book, Cultivating Joy and Wonder. (www.shelburnefarms.org)

STORY TIME: OWL BABIES

The Witch reads Owl Babies twice a day. We have the large format story book and three beany baby owls. She hands out the owls and gathers everyone around.

We started reading Owl Babies at Halloween one year for the little children who were too scared of Spooky Wooky.

We had a not-so 'Scary Tree' in the lobby which was a few twigs that had a sensor that moved around and a really friendly blow-up vampire that popped out of a coffin, but we soon realised even these were too terrifying for small people. The little ones love the Owl Baby story. It is not a long story, but they have so much to talk about, so it takes a while! It's a big hit every year.

NATURE QUIZZES

At every seasonal event, we have a new indoor nature themed quiz. This keeps children occupied while they are waiting to do an activity or if it is raining. We pick a relevant animal. For example, at Halloween we have had owls, bats, spiders, worms, etc.

How to Run an Indoor Quiz

Hand out the question sheets (you will also need a bunch of pens). Print out the answers to the quiz in large font accompanied by an image of the animal. Laminate them and stick them up with blue-tack in sneaky places.

The first answer is in view of reception, so the member of staff can point to the image of the animal on the answer and say, "Find the other answers around the inside play area and come back to me for your prize." The answers are dotted around the play area, so they must fill in the correct answer under the question.

They are rewarded with a chocolate coin when they return.

Quiz Example

Bat Quiz

1. How long do bats live?
2. How do bats find their food?
3. How many mosquitos can bats eat in one hour?
4. How fast do bats fly?
5. How many pups (babies) do they have in a year?
6. How large is the wingspan of the biggest bat?
7. How do bats sleep?
8. Do bats hibernate (sleep through winter)?
9. What is the name of the world's smallest bat?
10. Why are bat droppings (guano) valuable?

TICK IT OFF

If you want to design a simple activity that relates to your own environment, draw up a simple tick list. Children LOVE ticking things off. Offer them a challenge with places to tick and they are off! For example, we have flower and Santa tick off lists. See 'Tick it Off' in the bonus material: http://www.naturemagic.ie/bonusmaterial

SEASONAL EVENTS

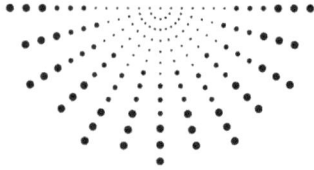

CHICKEN RUN AT EASTER

We designed this activity to avoid a traditional Easter egg hunt where the biggest, fastest kids get all the eggs and the little ones end up crying!

Someone dresses up as a giant chicken!

The chicken comes out at various times and clucks around. Each child on the hunt buys a stamp card with six sections. This way, we can manage the number on each hunt. (30 children max on each chicken run). It also means we can manage the number of eggs handed out. You only get the egg if you hand your card to the chicken (this avoids parents begging for extra eggs!).

The chicken needs a leader with a megaphone as well as a bodyguard! The leader gathers the group and explains that there are different activity stations around the nature walk. After each activity, a child will get a stamp on his card.

The leader also tells the group to warn the chicken if they see the fox.

The activities along the run are:

- Spud and spoon.
- Scream (burns off energy).
- The Birdie Dance.
- Hopscotch.
- Sack Race (there is a parents' round of the sack race which is hilarious but they can injure themselves…!).
- Skip.

The fox pops up in the rocks and at various places, causing all the children to scream, "Fox, fox!" And the chicken gets flustered and starts clucking and running faster.

The vital ingredient is to have a confident chicken, someone up for clucking and scratching and squawking and then able to get broody and lay eggs! Not everyone has these skills, so have auditions before you choose your chicken!

SPOOKY WOOKY AT HALLOWEEN

For the first five years, we put on an elaborate Halloween show: Spooky Wooky. Although there were many elements that were brilliant and hilarious, we now realise it was WAY too ambitious. We give this idea a thumbs down!

We turned the barn into a haunted house with corridors, populated with spooky people. Grace, aged 8, was a zombie in a cage who had to grab people by the ankles; Biddy the Witch (based on the legend of Biddy Early) gave a tour of the graveyard where she had buried her husbands; paintings fell down and clowns popped out; water squirted out of ancestors' eyes… Dracula played a creepy old piano and a poor soldier was melting in a barrel of acid. There were lights and music and spiders falling on people's heads. Eventually, Biddy got you to the cauldron where you made a spell and tossed in rats and fingers into the dry ice smoke. Every year, Biddy had a new dead husband!

It was a huge operation to build and run. People loved it; but the children were so scared, I think we traumatised half the neighbourhood. (Scooby Doo was there to whisk you out if you panicked!)

When other haunted houses started popping up around the country, we called it a day and now just do a few of the successful activities.

Roasted Marshmallows/Smores

Who doesn't love them?

Have a responsible person handing them out so they are not too hot and put it between chocolate biscuits (not on skewers).

Biddy the Witch

She is now a nice witch and pops in to say hello and play games with kids and reads the Owl Baby story to the little ones—a VERY popular event.

The witch needs a good costume and green makeup.

Monster Beanbag Throw

Throw the beanbags into the monster's mouth.

Turnip Bowling

Fill some old plastic bottles with coloured water and use a turnip as the bowling ball.

Spooky Dancing

Last year, we had a Halloween workout running on the TV with pumpkin Pamela, "ride the witch's broom," etc. This was popular, but it exhausted Biddy as she had to do it with the kids about 20 times a day!

Face Painting

You can't go wrong here, especially at Halloween. We also offer animal designs, ladybirds, butterflies and fairies.

SANTA'S ON HIS WAY

Having visited a few Santa experiences over the years, we knew exactly what we didn't want: rubbishy plastic toys, fake beards, cold and miserable queues, expensive photographs and bored, grumpy Santas.

It had to be magic for the child and the parent.

When we first opened, it was during a recession and a lot of fathers were working abroad. They just came home for Christmas; that photo of the family together can very important.

Our top tips for running a Santa experience are:

- Find a magical Santa with a real beard! Make sure he genuinely loves children (all staff have to be Garda vetted, so do this in time).
- Find a funny Mrs Claus. Nobody wants a boring grandma. Our Mrs Claus loves Santa and is a wonderful host at her kitchen table, but she is a feminist and is notoriously bad at housework and cooking! She brings a pie out of her oven and burns iron shaped holes in Santa's boxers (that are hanging over the fire!).
- Audition your elves and train them well. Give them a script and make sure they are always chatty and engaging.
- Get the booking system right. It must be online and capture all the details. Do not take bookings over the phone—it is way too labour-intensive and confusing.
- No plastic toys. We have now had 7 years of Santa at Burren Nature Sanctuary with no plastic toys. Who wants to add to that mountain? Start looking as early as January! And bargain with the toy producers, look out for Jenga toys, Timber fairy doors, Timber puzzles and craft kits like paint clay frames or fridge magnets. There are low margins on Santa events, so keep the costs down.
- You cannot charge full price for babies, but they will still want a toy! Have soft baby balls or little teddies for babies under two years.
- We try to create a natural grotto; we bring in off cuts of branches and spray them white and have as many retro items as possible.
- Theme it around a different animal every year. Children have an insatiable thirst for knowledge about animals, insects and birds.
- A big train set is essential. It's tricky to keep it going. We have set ours up so there is a large button for children to push to start it. It is very disappointing for children to look at trains but not be able to play with them.
- Christmas stables: we normally bring in the alpaca, a donkey and maybe a goat. We decorate the shed with lights and stars, and there is a snow machine (run by the animal handler).
- Nativity Dress up: costumes for the nativity characters are

hanging on a rail and Cookie the donkey is in the stable for a photo shoot. Some families have a lot of fun with this.

The running order of the Santa experience is:

- Do the nature themed quiz and win a chocolate coin (and play in the soft play while you wait).
- The chief elf brings the groups into the Santa experience. There you can enjoy games, play with the train and have a cookie/tray bake with Mrs Claus. She also provides tea, coffee, hot chocolate and mulled wine for the adults. (We only realised after seven years that we are the only Santa experience offering real mulled wine; all the others are packet mixed and non-alcoholic!)
- The elf makes sure Santa knows which family is coming in next and he has the information about the children (captured at booking, for example, 'Tom wants a Barbie', 'Isabell loves tractors', etc.)
- Santa talks to the children individually and never rushes them. He gives them all a Christmas decoration as a memento.
- The photographer elf takes the photo. This used to be emailed, but people mostly do it on their own phones now (to avoid the Santa photo scam where you have to buy it at an inflated price). Make sure your backdrop is respectable.
- The elf takes the family to the elf workshop where the children can choose which toy they want. This avoids the howls of despair when they open their presents and get something they don't like.
- The elf shows them out to the Christmas stables where they can meet the rabbits, hold a guinea pig, do the nativity dress up and get snowed on!

Note: parents are very frazzled at Christmas with all the stress. Expect them to be demanding and make sure everything runs smoothly.

Try to keep a smile plastered to your face!

Mrs Claus will be under a lot of stress keeping people entertained while they are waiting, so make sure she has a superlative sense of humour and a lot of patience!

Last year, our Mrs Claus had an exercise bike and made all the dads get up on it to get healthy! Surprisingly, dads seem happy to do anything for Mrs Claus! She also had a juicer and was making green juices for Santa.

Nature Magic

PART THREE
OPEN A NATURE SANCTUARY

NATURE LOVER INTERVIEW: COSTAS CHRIST

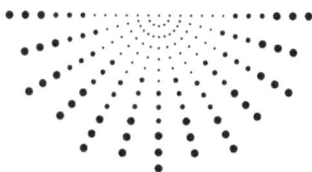

If I had that magic wand, I would put half of our planet under full protection.

COSTAS CHRIST IS AN ECOTOURISM PIONEER
AND FORMER SENIOR ADVISOR FOR SUSTAINABILITY AT NATIONAL GEOGRAPHIC.

HOW DID YOU BECOME A NATURE LOVER?

I grew up—as we used to say in my neighbourhood—a hard, scrabble kid. I was raised by a single mother who was a waitress for forty-six years. That is how we survived as a family of four.

My childhood was an unpredictable world.

As a kid, I felt that people were unpredictable and often, untrustworthy. I'm not referring to my late mother—bless her heart—because she was the star of my life and instilled in me the virtues of honesty and integrity that I believe in today.

But during those early and sometimes turbulent years, I always found nature to be a refuge for me. As a boy, I used to go into the woods and pretend that I could have conversations with animals who were my friends. I was just one of those kids who was born a nature lover. And while the outside world sometimes seemed crazy, I found stability in nature.

As I grew older, I felt that I also had a responsibility to give back to nature. That had brought me comfort as a child (and it still does).

So, whereas nature helped me as a youth, as I grew older, I wanted to work to help nature.

WHAT IS YOUR FAVOURITE PLANT OR ANIMAL?

I always had a passion for sea turtles.

I grew up on a barrier island off the coast of New Jersey. And when I was young, a very large sea turtle once washed up on the beach, dead. It was huge; in my eyes it seemed to be the size of a Volkswagen Beetle car! I stared at it for a few hours, fascinated.

Growing up, I learned more about sea turtles and giant tortoises; just think of it, an animal that can easily live more than a hundred years. You look at the face of these unique creatures, as I have done while working to protect sea turtle nesting sites, and there is a sense of wisdom. Since you asked about my favourite plants, well, you're speaking to a blueberry farmer and I'd be lying to you if I didn't tell you that blueberries are my absolute favourite!

DO YOU FEEL SPIRITUALLY CONNECTED TO NATURE?

I find nature in so many respects to be a spiritual experience.

Trees… I mean, it is a euphemism, but there is the sense, when walking into a forest, of entering nature's cathedral. I live in Maine, a state that is part of what they call the Great

North Woods. I'm surrounded by vast spruce forests and entering there is truly like walking into a majestic cathedral.

Growing up on a barrier island, the ocean has also been a source of deeper connection for me. I learned to surf when I was twelve years old and surfing became my passion. Sitting in the water, you cannot help but think about the energy that moves through the waves. Capturing that energy and riding that wave can be a very spiritual experience, something that surfers often talk about.

WHAT CAN PEOPLE DO TO HELP NATURE?

One of them would simply be to reflect on our connection with nature. When we protect nature, we protect ourselves. When we heal nature, we heal ourselves.

On a more pragmatic level, a lot of my work is connected to the travel industry. I was involved with helping to develop the early concepts of ecotourism, which was defined by a small group of about 12 of us over thirty years ago as, "Responsible travel to natural areas that protect nature and sustains the wellbeing of local communities."

We knew from experience that when local communities see their own livelihoods directly connected to protecting nature, such as through ecotourism-related jobs, then they become active partners and allies in supporting conservation.

In terms of a positive action when we travel, perhaps most important is to understand that our travel choices can make a huge difference. We need to support those travel companies committed to the principles of ecotourism. For example, the local companies that are part of the Burren Ecotourism Network in Ireland. As a traveller, wherever you go, seek out and give your business to those hotels, tour operators and destinations that are also working to protect the environment, embrace cultural heritage and benefit local communities.

IF YOU HAD A MAGIC WAND, WHAT IS THE ONE THING YOU WOULD YOU DO TO HELP THE PLANET?

> I would put half of our planet under complete protection. There is a very famous ecologist, they call him the grandfather of biodiversity, E.O. Wilson. Some people say he first came up with the term biodiversity—meaning life on earth, all the living species.
>
> E.O. Wilson said that for our planet to survive, we have to protect at least 50 per cent of our oceans, our lakes, our streams, our forests, our savannas, etc.
>
> He calls this the Half Earth Initiative. So, I if had my magic wand I would wave it right now and put half the Earth under protection.

Visit www.naturemagic.ie (or search 'Nature Magic' on all major podcast platforms to listen to the full interview).

HOW TO OPEN A NATURE SANCTUARY

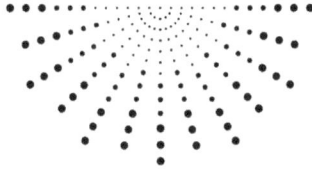

INTRODUCTION

Do You want to Open Your Own Nature Centre?

It is a vast undertaking to open a nature centre. We went from planning, to building, to winning awards in three years.

If we had known everything we know now, it would have saved so much time, hair pulling, expenses and exhaustion. There are so many elements you have to get right: planning, designing, launching, marketing and smooth operating.

It is a huge topic, but also very exciting!

I would like to give you a warm congratulation if you are embarking on this road. You deserve a big pat on the back. If you've already started but feeling discouraged, read on. The inspiration you need to continue may be hiding on the next page.

These are tips I feel would have been useful to us when we embarked on the journey of opening and successfully running a nature sanctuary.

If you need more in-depth advice on a specific topic, please feel free to contact me. I'd love to help you work it out.

Tips Before You Open the Doors

Let's say you already have a site to build on. The first step is to sit on the site and visualise exactly what you want to create. The next step is to design and develop your sanctuary! The following aspects need your careful consideration:

INSURANCE

Insurance is the stuff of nightmares in the 'adult world'.

The infamous 'compensation culture' in Ireland is now closing businesses daily. A claimant conveniently only has to pay a solicitor to write one letter and many lawyers even work on a 'no win, no fee' basis.

Sadly, people feel entitled to receive compensation for ridiculous amounts if they hurt themselves on your property. Judges encourage this sense of entitlement when they grant huge settlements. People are ignorant about where the money to pay for such claims comes from. They think, "Oh, the insurance will pay for it. It's not like we are getting it off the business." They never think about the hike in premiums and many local people who depend on the businesses for their livelihoods.

Burren Nature Sanctuary has had one claim since opening (touch wood). The person and a group of special-needs children were granted free access to the sanctuary, as a community gesture from our side. She hurt herself and sued for compensation. The insurance company ruled the claim invalid, as we were not negligent in any way.

Although we did not need to pay the claim, our insurance premium tripled. Staff members lost their jobs, and the claim remained on the claims statement for five years, meaning no other insurance company would give us a quote. We nearly had to close our doors. And our hefty premium never reduced...

CHOOSE A NAME

One of the very first decisions you will have to make is choosing the name of your nature centre. If you are having trouble with this, consider naming it '[insert your local area here] Nature Sanctuary'. Next, register it with your local Companies Office. If your chosen name is not available, try different versions of it.

We eventually decided on Cloonasee Ltd (Meadow of the Fairies) T/A (trading as) Burren Nature Sanctuary.

BUY YOUR DOMAIN NAME

As soon as you have confirmation that your company name has been registered successfully, you need to get on the internet and buy a domain name for your website.

Ideally, get it according to your own country prefix. For example, we own Burrennaturesanctuary.ie and Bns.ie. But we also hold other versions, so other people can't buy and use them.

In our case, the 'ie' is the country prefix for Ireland.

WRITE YOUR VISION AND MISSION STATEMENTS

It's essential to take your time while formulating your vision. It will be dynamic and change over the years. But if you don't have an obvious purpose about what you want to achieve from the start, you won't have goals to work towards, and people will not support your business.

Think of your mission statement as a roadmap to achieve your vision. It will contain smaller, short-to-medium-term goals aimed at helping you reach your vision. See our Vision and Mission Statement in the Introduction and note how our mission statement helps us work towards our vision. To formulate your mission statement, break your vision statement into small 'chunks' and then, for each chunk, ask yourself, "How am I going to achieve that goal?" The answers will become a part of your mission statement.

DESIGN YOUR LOGO

We really struggled with our logo and changed it after five years.

It's a lot easier to get a great logo these days with online platforms like 99Designs around. Your logo should be minimal and clear. For a nature centre I suggest a green palette, as your logo should say 'nature' at first glance.

Our logo is now a garland silhouette of native Burren flowers. It is both beautiful and compelling. You could copy this idea with your native flora.

Your logo is often the first thing people will see and subsequently think of when your business comes to mind. It is well worth the investment to have a proper, professional logo designed.

THE ENTRANCE TO YOUR NATURE CENTRE

The entrance must be broad enough for a car to turn around in if they decide not to drive in. Consider the visitor numbers you expect and make sure there is enough room for a coach to turn in comfortably. It should be rustic with stone and timber and no gaudy, colourful signs.

We hope to have new entrance signage made. Our large aluminium sign blew down in a storm, but we didn't mind too much as it had the old logo on it. We will invest in a beautiful, professional, rustic entrance sign as soon as funds allow it.

Ideally, your sign should have a changeable slot for opening hours.

BUILDING DESIGN

The most important thing to consider is visitor flow. Planning is crucial. Because Burren Nature Sanctuary started in a converted barn and wasn't a new build, and we had no experience or advice, we have struggled immensely with this issue and had to implement creative measures to ensure a smooth and enjoyable experience for our visitors. It is much easier to get right when building from scratch.

WELCOME AREA

It should shout 'Welcome, come in, this is where you go!' Have clear signage in the reception area with prices and a blackboard showing the times of different activities. People should walk in and out through the shop.

OUTDOORS

The outdoor area is unique to every nature centre. Go to similar attractions that you enjoy visiting to gather ideas for your own paths, shelters, animal paddocks, exhibits, fencing, signage and nature interpretation, etc. I am happy to advise on design and layout of any aspect of your facility.

ADVENTURE PLAY DESIGN

We have an indoor soft play area and an outdoor adventure playground.

The key to designing a play area is that the children are safely 'penned' in. At Burren Nature Sanctuary, a green 6,5-foot (2 m) high wire v-mesh fence surrounds the play area (and the only entrance/exit is through a Maglock gate).

We always require parental supervision, but parents and staff can relax knowing that if a child gets 'lost', they cannot escape down the nature walk or into the car park. This is a great relief for parents and especially carers of hyperactive children. They may even sit down and enjoy a cup of coffee! When we opened the first couple that brought their toddler to play said to us, "That is the first time we have had a conversation in over a year."

It is important to make play areas adult-friendly. The children will be happy anyway, but there is nothing worse than being penned into a smelly, ugly soft play area with bright fluorescent lighting smelling of old burnt chip oil.

Adults want somewhere nice to sit, good Wi-Fi to check emails and a stellar cup of coffee or tea. It also makes life easy for mums and dads to provide reasonably priced healthy children's food and delicious options for adults.

We cook hidden vegetable tomato sauce for pasta and pizza. 99% of children like pasta. We do not sell any sweets, bars or cans of sugary fizzy drinks. This is a conscious decision on our part as it would be a very good income stream. We sell home-baked cupcakes, brownies and cookies as treats. From experience, when young children go to a soft play centre there are always sweets and they become obsessed with them…

"When can we have sweeties?"

Normally the answer is, "You can have something when we leave."

They reply, "Can we go now then?"

Nature Quest Indoor Play Design

Design your soft play in tones of blues and greens for a calming effect. Choose green for the slide so it looks like a grassy hill.

We designed our indoor soft play to incorporate nature into the experience. The PVC (washable!) kids' educational panels are displayed along the outside net and each panel requires an action. All the children read the signs when taking rests from play.

The Pollination Game

We do not have a ball pool (not hygienic!) but we use the yellow pool balls to play this game.

At the top of the play construction are two options: a flower or a honey pot. Children can decide whether the yellow 'pollen' balls will go to pollinate another flower or are taken back to the hive for food. They post the balls in the hole and they slide down the two transparent tubes to large pockets at the bottom of the entrance, on the lower level. Having run frantically up and posted several balls, descending the four-lane wavy 'hill' slide, they can then see 'who' has won—the flower or the beehive. This game is suitable for teams.

The Hill Slide

There is no point investing money in a soft play area (ours cost €50k) unless you include a four-lane wavy slide. This provides a circular route and is the means to burn off megawatts of extra energy. Before the children are old enough to manage the slide themselves, parents are forced to crawl up on their knees to take their kids up. By the tenth time, the parents are puce and sweating. The children are shouting, "Again, again!"

Burren Flower Hunt

Bee orchids, burnet roses, bird's-foot trefoil, pyramidal orchids, maidenhair ferns and mountain avens are stitched in different places of the play area. The corresponding panel explains to children where these flowers originated from with a map and asks them to see how many they can find of each species in the soft play area. The answers are under PVC flaps on the sign.

Moth Hunt

The green Burren moth and the burnet moth are also stitched on for children to find.

Fossil Hunt

The last PVC panel explains fossils and asks the children to find them on the rocks in the playground.

Name the Parts of the Flower

See if you can name the parts of the flower. The answers are hidden under little PVC flaps.

Photosynthesis

(For school tours.) Use the atom kit to make the molecules that plants produce and we breathe.

Baby Farm Play

We designed the baby play section as a barn with low, movable stall walls and rocking animals—a pig, a sheep and a cow. There is also a small climbing hill and a wobbly farmer. The wobbly farmer is often subject to attacks by ten-year-old boys (for unknown reasons), who have to be removed from the baby area!

Baby Puzzles

There is a Velcro (hook-and-looped fastened) farm puzzle in the play area and a 3D shape posting puzzle with a bee, a ladybird and a flower that fits into a piece of wall art.

Tree House

Above the baby area, looking out at the tables where the adults sit, is a tree house. We designed this on the outside of the play structure with windows to look out of and wave at your adults. Our wonderful local artist, Susan Meaney, (www.burrenflowerfairies.com) designed a tree with

Burren flower fairies. It goes from the ground up to meet the tree house. This also helps to soften the look of the structure.

The area also includes the following for fun:

Talking tubes (cones with tubes to talk to people in a different area).

Indoor Zip (a buoy on a rope that you can sit on and ride along).

Climbing Wall (with bird silhouette cut outs).

Stepping Stones (designed as tree stumps).

Hoops (to swing off).

A Bee Spinner (you sit on a little disc and can spin the pole until you get dizzy or fall off. Padded walls surround it, of course!).

Spider's Web (to climb through).

Even though the play area is small, we managed to fit all the exciting play elements in.

Every evening, the play area gets fully sanitised, and every morning we undertake a full inspection, signing off every element of the play area before any children are let in. Each year, we have an official, full 'ROSPA' inspection of both play areas. They inspect our daily reports and every inch of the play areas. They rate each element for risk and may require us to take remedial actions, for example, to replace pads or netting that is getting worn or add more sand to the landing area of the slide, etc.

The Burren Challenge Adventure Playground

The outdoor play ground is divided into two sections:

The Sand Pit

The first section of the outside playground is for children under 5 and is a large sand pit. It includes a sand construction site station with pulleys to lift sand, a slide and a ride-on digger. Amazing construction projects have been accomplished in this area. We often find all the children working together, building towering sand castles and digging immense holes (that have to be filled in every morning!).

Once, an adult was witnessed (by another parent through the window) hitting a child with a plastic shovel. (We are all trained in the 'Children First-National Guidelines for the protection and Welfare of children' and have policies in place if any sort of abuse is witnessed or disclosed by a child on the premises.) We caught the entire episode on CCTV (We have 25 cameras throughout the premises) and reported the individual. It wasn't even his child… He wasn't punished. But let's hope if he is ever involved in another incident, this file will resurface. So, you can see that children bring several challenges to the premises that people may not consider, including high insurance premiums for play areas.

Adventure Playground

We designed the main outdoor adventure playground as an assault course around the large Tower Slide central play element and the Mobilis. There is also a set of swings in the top far corner. No playground is complete without swings!

The Tower Slide is a conical, three level climbing frame with metal tubular slide.

The Mobilis is a rotating see-saw that also goes up and down. It is surrounded by a rope fence with a strict policy of only two people inside the rope at one time!

The Assault course: is mostly timber (rather than metal) to add to the rustic feel of the whole site. The aim is to go from start to finish without touching the ground. We have set limestone stepping stones with names of Burren mountains into the grass: Oughtmama, Turlough Hill, Mullaghmore, Slieve Carron and Abbey Hill.

- Start at the balance pole.
- Then the stepping stones.
- Zig zag balance.
- Then the hoops.
- Then the rope hang.
- Then the net crawl.
- Then the monkey bars.
- Then the net and rope swing.

- Then the main event: the 164-foot (50 m), double track zip line!

There are certain rules in the playground and these are also stated on the sign-in sheet at reception which include:

- Only one person at a time on each zip wire ride. You must sit down, be over 8, and hold on with two hands.
- At busy times, we have supervisors in the playgrounds to discourage horseplay. The toughest age group to monitor is boys exhibiting bravado between ages 8 and 12. (And naughty adults.)

The playgrounds may be the first draw for children, but then they all also want to meet the animals and nearly every family puts on their coats and does the mile-long (1,6 km) nature walk. The walk seems an achievable option for frazzled parents who don't want to make decisions on their day off. There are signs to follow and it is a circular route, so you can't get lost!

The Bird Box Treasure Hunt and **The Fairy Tale Fairy Trail** stories help parents to encourage children around the walk. (https://www.naturemagic.ie/bonusmaterial)

PRE-PLANNING PERMISSION

As soon as you have draft plants, set up a pre-planning meeting.

Try to get the council's support from the beginning. Being able to address their concerns about traffic, your entrance, the build etc. is key to your success.

GRANTS

Look to your local authority for financial support.

We got a LEADER grant from our local rural development office for €200,000. We wouldn't have been able to open Burren Nature Sanctuary without this.

Grant applications are a challenge!

Ours ended up taking up two large archive boxes. Each item, from the smallest nut and bolt, had to have three quotes for it…

It's lovely to get the money, but you certainly work for it.

⁂

TIPS AFTER YOU OPEN THE DOORS

ADMISSIONS, TICKET PRICES AND MEMBERSHIPS

Research your competitors. You don't have to be cheaper, but you need to be in the same price range. Complaints are normally about price.

THE GIFT SHOP

Choose a few lines of local artisan products. Have some small items under €5 near the till.

You can set up an online shop at a later stage.

If you produce something on site, this is ideal as it will have a more significant margin than re-selling another person's product(s).

CAFÉ

Keep your menu small, and use local ingredients as much as possible.

If you have an excellent year-round footfall, the café menu can expand but BEWARE: the café can be a big cash burn. You need to buy supplies, pay staff (very expensive) and the profit margin on food is tiny.

Against that, people will be more likely to visit you if you have a good food offering. From our daily, weekly and yearly sales pie chart, 50% of our income comes from admissions, and 50% comes from café food and gifts.

Top Tip: use coaster buzzers (a restaurant calling system) for orders. These are ideal for a larger site. You can then call your customer when the food is ready to collect (it works within a 1,2 mile [2 km] range). This avoids a waiting service, hungry people hanging around—stressing the team asking whether their food is ready—and staff wandering around with sandwiches, looking for people who have moved tables or gone out to the playground.

STAFF

Staff will make or break your business.

Work alongside your staff and use only positive feedback. Communicate any problems clearly and discreetly. With well-trained staff, you will rarely have problems.

Get clear contracts drawn up with trial periods. If someone is really not working out, let them go during the trial period. This is tough but kinder on both sides in the long run.

Staffing is the most challenging area of running a business and the most expensive. For the business to break even, staff costs must hover below 30% of your turnover.

This is easy to achieve in high season when turnovers are typically large, but it is still easy to spend too much on staff.

It's a lot harder to keep staff costs within the 30% margin during the low season. Plan and determine exactly where staff are essential and how many hours and days you will need them for at each station.

Other things to consider in low season to keep staff costs as low as possible:

- What are your opening hours?
- Are you going to hire seasonally or attempt to employ competent staff permanently? The former is very stressful, you may be in a bidding war with other seasonal businesses, but the latter is costly.

It is tough to get reliable staff. If you have a good member of the team, treat them with respect and allow flexibility for family issues.

Never be late with payroll.

We employ a lot of young people, and they are excellent. But they need to be trained and managed well. We expect as much from them as we would from mature adults. But we also assume they will mess up at some stage and that is ok, too. Second-year young people are usually so brilliant and confident, they could run the show!

Staff Hand Book

The staff handbook should contain updated training material and company policies. Each staff member must read and sign it.

Staff Training

To ensure our staff are knowledgeable about the sanctuary and its operations, and the Geopark Code of Practice, they complete a pre-training survey. The survey tells us what they know (or don't know).

Afterwards, they watch a training slideshow and receive a training sheet to work through.

Two weeks after the training, they have a role play test with a pretend difficult customer!

See our Staff Training Factsheet here:

https://www.naturemagic.ie/bonusmaterial

COMPLIANCE

You will need to comply with many regulations and will end up with pages of company policies in your staff handbook: from building regulations to health and safety, food safety, play safety, etc.

WEBSITE

You will need excellent, large images for your website. The right image is everything.

Because we grew organically and our centre is a converted barn, we are still struggling to get that FOMO ('fear of missing out') image that gets people behind the wheel to come and visit…

To work out your website, and save on design time and costs, get some sheets of paper and write on each what you want on your website pages; and draw little boxes to work out the layout.

Gather your content and write the text you need on the website. Ask your web designer to include booking systems for parties and groups and an online shop.

ACCOUNTS

Do not try to handle your accounting on your own. Pay a professional accountant from day one, it will save you money (and prevent premature grey hair).

ACCOMMODATION

If you have accommodation available on-site, it is an excellent complement to the business. Alternatively, try to befriend accommodation businesses in the area and negotiate with them to accommodate your guests, especially for booked-in groups.

MARKETING

This is a big topic, but options include:

- Leaflet distribution (expensive and not eco-friendly).
- Social media and newsletters (time-consuming and need to be consistent).
- Marketing groups (expensive).
- Networking (extremely time consuming and immersed in local politics).
- Speed dating with tour companies (expensive to attend but good value in the long run).
- Forging relationships with relevant journalists (time-consuming but effective).
- Awards (difficult to win, but it provides good media coverage).

Burren Nature Sanctuary has won:

- The Scull Enterprise Award for Arts and Tourism 2013.
- The Burren and Cliffs of Moher Geopark Award for Leave No Trace 2014.
- The Burren and Cliffs of Moher Geopark Award for Interpretation 2015.
- Sustainable Tourism Business Award, Silver 2016.
- Eden Award for Gastronomy 2017 (as part of the Burren Food Trail).
- Bank of Ireland friendly Business awards 2018.
- Business for Biodiversity Award, 2nd place (to Bord Gais), at the National Biodiversity conference 2019.

TOUR COMPANIES

Working with tour companies is a topic in itself.

Pitching and selling your attraction and experience is a finely wrought craft. You will need training and practice!

FUNDRAISING INITIATIVES

Asking the community to support you through fundraising initiatives may help you cover costs and create environmental awareness amongst locals.

Using a little creativity, you can get people excited about helping. For example, during the lockdown period of 2020, we started an 'adoption' system for our animals.

We took beautiful portrait shots of all the farm animals and sat down with the Burren Nature Sanctuary team. We designed a certificate which each included group admittance for five people to the sanctuary and a personal introduction to the 'adopted' animal.

We made the 'adoptions' available through our online shop and shared it on social media. The people who adopted animals helped us immensely to raise much needed funds. As soon as the lockdown lifted, they came to the sanctuary to meet their adopted animals.

Another of our initiatives, the Native Tree Project, is also available through our online shop and helps us raise funds to plant trees. If you would like inspiration for creating your own fundraising initiatives, visit the following links to see how we did it on our website:

https://www.burrennaturesanctuary.ie/shop/

https://www.burrennaturesanctuary.ie/product/donate-e10/

TIPS FOR HANDLING REVIEWS

Social proof (known as 'word of mouth' before the social media era) is essential for your business. Reviews tell people who are considering visiting your nature sanctuary whether it is worthwhile going.

General Tips

- Respond to all reviews.
- Respond to positive reviews with enthusiasm.
- Never engage in an argument or respond negatively or defensively to bad reviews.
- Ask a negative reviewer to email you with more details.
- Complaints are often about the price. Most people don't like to pay! Have a standard reply that lists everything you offer at your site and thank them for their feedback.
- Take any negative feedback seriously and address it instantly.
- Print out positive reviews for staff members.

TripAdvisor Tips

- Make sure your business profile is up to date and professional.
- Add your own photographs.

This is our latest TripAdvisor review:

What an incredible hidden gem! (5 stars)

Travelling to the Burren last weekend, we stopped at the Burren Nature Sanctuary in Kinvara for the first time. It is a wonderful and magical place to visit.

We began with a delicious lunch (soup, cheese scone, salad, nettle cheese and chutney) and excellent coffee in the cafe. Then we walked the lovely trail that took in ancient woodland, a famine village, and a fairy wood (enchanting, and we are four mature adults!). We returned to the cafe for more coffee and browsed the gift shop.

The staff on the day, Eimear, was really helpful and informative, especially about the famine village, as she is an archaeologist.

I couldn't recommend this place highly enough. There is also lots for children to do (indoor and outdoor playground and activities) but these were not for us. We will definitely return on our regular weekends in the Burren, and this time we will try the cakes too.

Facebook Tips

Encourage people to check-in on Facebook when they visit your sanctuary, on café menus or on your Wi-Fi sign-in page.

This is our latest Facebook review:

We had the most wonderful time at the Burren Santa Experience this year. My three kids—aged 9, 7 and 4—were just mesmerised from start to finish. Really well done between Ms Claus's Parlour and Santa (who both were so amazing, by the way) down to meeting all the animals in the barn after.

Thoroughly enjoyed it and would highly recommend.

#creatingmemories #bestsantaever

Google Tips

Google reviews are the most important. Whenever someone searches for your website using Google, these reviews will pop up. A good rating on Google is valuable.

These are our recent reviews on Google

Welcome back! Great visit this morning, first after the Covid-19 restrictions. Very good measures in place. Can't beat the facilities. Our kids love the indoor Softplay area. Keep up the great work. See you next week.

(5 stars)

Very enjoyable morning spent here with family (incl. 5-year-old) and the welcome from the staff could not have been nicer. She asked so many questions and talked to our son, making him really excited about the fairy trail and outdoor play area.

Lucky enough to catch Emilia, the pig, for a walk on our way around the paths, ending with a good play outdoors in the large kids play area. Then inside to the soft play, while we both had a lovely coffee. Definitely recommend it.

(5 stars)

Dealing with Complaints On-site

If you have a complaint on-site, take out the customer complaint form and ask the customer to fill it in. Usually, they won't want to as they just want to rant at a staff member or get a discount. It is very unpleasant when staff are put in this situation, but if they have the form to hand over it defuses the situation.

We only had one challenging complaint: in our first year, we hosted a christening party. The whole café room was set up, and everyone seemed to enjoy their food. When it had all been served, the staff high-fived each other in the kitchen—everything had gone so smoothly. But after the party, the christened child's mum came to the counter and was furious. She complained about the food and service and threatened to write to the local paper. We ended up giving her half her money back and some vouchers.

Two years later, her ex-husband paid us a visit. He apologised and said that the food had been lovely, and everything was perfect. He was really sorry about the whole thing.

So, you see, you can do your best to please people, but you can't please all the people all the time!

MAGICAL STORY
THE HOLY FIRE

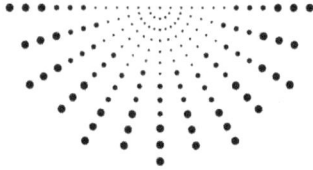

Before we opened the doors of the Burren Nature Sanctuary, my neighbour dragged me to a clairvoyant life coach who was visiting the area.

At the time, we were constructing the nature walk. Not an easy task. Bob and David had a mini digger and a mini dumper; they were working hard to make a path through 50 acres of shattered limestone pavement, untouched by the hand of man (for I don't know how many millennia).

They had to wiggle through the large, scattered boulders. They broke through whitethorn barricades and revealed nature's secret garden: wildflowers, climbing roses, trees, bushes, foxes, badgers—all flourishing and untouched by human hands.

At the start of the walk, there were what looked like a few dilapidated sheds. But it soon became clear there were more than we thought, and it was actually a group of ancient dwellings.

We discovered a round field that was probably used as a pen for cattle. As we made the path, small mounds of oyster shells were uncovered! These were middens (old dumps for domestic waste), proving that humans had occupied the area in the past.

So, we went along to see Reshmi, and she shuffled cards. I asked her if the land was happy to become a visitor centre, open to the public.

"Yes, all the spirits of the land are happy about it, but you must make a holy fire for the souls of the departed. It should be in the round field. (She had never visited the farm.) You must make the fire of four types of wood found on the farm: oak, hawthorn, hazel and t-tree."

"Tea tree?" I said, "That's Australian…"

"Yes, definitely the t-tree. That's the message."

This left me puzzled for months.

One day, I was researching the ancient Ogham language of Ireland. Each letter has the name of a tree or plant. The tree for the letter "t" was the holly. We had holly trees on the farm.

Holly was the holy tree. It was a holy fire.

'That must be it!' I thought. I put on my coat and hat and grabbed the matches.

Roy asked me, "Where are you going?"

"I have to light a holy fire for the souls of the departed at the round field."

He rolled his eyes.

CONCLUSION

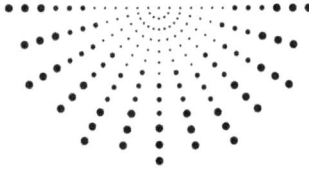

I hope you enjoyed reading this book and that it proves to be a useful guide and inspiration for helping to bring people back to nature.

Please feel free to contact me if anything is unclear or erroneous, or if you have any further questions.

https://www.naturemagic.ie/contact

Good luck with all your endeavours fighting the climate and biodiversity crises.

Let's create a nation of nature lovers!

ACKNOWLEDGEMENTS

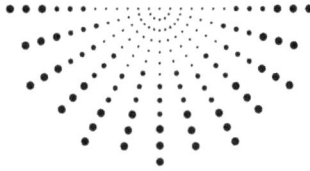

To all the amazing staff who have worked at Burren Nature Sanctuary over the last seven years, thank you! There are too many to list, but you know who you are and you are all brilliant.

Special thanks to:

- **Helen Dagg Reynolds** for being Mrs Claus, Biddy the Witch, The Easter Chicken, The Fairy Pig Walker and for never failing to be there for us.
- **Daniel Reynolds** for maintaining the Fairy Wishing Well.
- Santa (alias **Paddy Neylon**) for being pure magic.
- **Eddie Dee** for his amazing work on the Burren Botany Bubble.
- **Bob and David Flaherty** for ongoing fun and help with the animals.
- **Anne Forde** for beautifully illustrating our map and fairy stories.
- **Kieran Whelan**, our builder, who encouraged us from the start.
- **Aiden Murray**, our plumber, especially for sorting out the flood at 7 AM on a Sunday morning!
- **Dr Noeleen Smyth** for being a great support with the botanical collection.
- **Tom Assheton** for business advice.
- **Mary Bermingham** (Roy's sister), thank you for your help, love and kindness.
- A very special thank you to the late, great **Joe McCormack**— our electrician, who never failed to have us all in fits of laughter.
- And not forgetting the **fairies**… Or the **animals**… And the **plants**…
- And thanks, **Roy Bermingham**! :-)

- And **Alice**, **Arthur**, **Lorna**, **Grace** and **Zara**. :-)
- Thank you to **Laura Peterson**, for her brilliant writers' group and coaching. (www.copythatpops.com)
- And the fabulous **Melissa Niemann** for editing Nature Magic! (https://bit.ly/melissaniemann)

APPENDICES

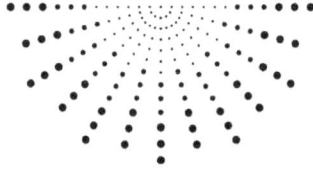

BONUS MATERIAL

Follow this link to access all the free bonus material mentioned in this book: http://www.naturemagic.ie/bonusmaterial. Contact Mary at https://www.naturemagic.ie/contact

PODCAST GUESTS' RECOMMENDED BOOKS

- *The One Straw Revolution by* Masanobu-Fukuoka

PATRICK MCCORMACK'S RECOMMENDED BOOKS:

- *So Shall We Reap* by Colin Tudge
- *Arctic Dreams* and *The Crow and the Weasel* by Barry Lopez
- *Anam Cara* by John O Donohue
- Check out the books by author John Moriarty.

GONÇALO SANTOS' RECOMMENDED BOOKS:

- *A field guide to the Orchids of Europe and the Mediterranean by* Rolf Kuhn, Henrik Pedersen and Phillip Cribb

- *Collins Wild Flower Guide by* Davide Streeter, C. Hart-Davies, A Hardcastle F. Cole and L. Harper
- *Ireland's Wild Orchids* by Brendan Sayers and Susan Sex
- *Wild Plants of the Burren and the Aran Islands* by Charles Nelson

CATHERINE FARREL'S RECOMMENDED BOOKS:

- *Gaia* by James Lovelock
- *The Web of Life* by Fritjof Capra
- *Our Once and Future Planet* by Paddy Woodworth
- *A Sand County Almanac* by Aldo Leopold
- *Swan: Poems and Prose Poems* by Mary Oliver
- Check out Catherine's novel, *The Easter Snow,* here: www.catherinewilkie.ie

LAURA POWERS' RECOMMENDED BOOKS:

- *Secret Life of Plants* by Peter Tompkins and Christopher Bird
- *Be Nice to Spiders* by Margaret Bloy Graham

JANE STOUT'S RECOMMENDED BOOKS:

- *The Selfish Gene* by Richard Dawkins
- *The Blind Watchmaker* by Richard Dawkins
- *The Wildflowers of Ireland* by Zoe Devlin
- *Bees of Britain and Ireland* by Steven Falk
- *The Solitary Bees* by Bryan Danforth et al
- *Extraordinary Insects* by Anne Sverdrup-Thygeson

EAMON RYAN'S RECOMMENDED BOOKS:

- *Gaia* by James Lovelock
- *Silent Spring* by Rachel Carson

- *The Sea Around Us* by Rachel Carson
- *Nostos* by John Moriarty
- *Limits to Growth* by Donella H. Meadows

PADRAIC FOGARTY'S RECOMMENDED BOOKS:

- *The Sea Around Us* by Rachel Carson
- *The Unnatural History of the Sea* by Callum Roberts

JOSHUA STYLES' RECOMMENDED BOOK:

- *Collins Photographic Wildflower Guide* by David Streeter

NOELEEN SMYTH'S RECOMMENDED BOOK:

- *The Song of the Dodo* by David Quammen

JONATHAN PORRITT'S RECOMMENDED BOOKS:

- *The Lorax* by Dr Suess
- *The Peregrine* by J.A. Baker (read by David Attenborough on Audible)

COSTAS CHRIST'S RECOMMENDED BOOKS:

- The Fisher Lad audio by Urashim Taro (https://historyofjapan. co.uk/wiki/urashima-taro-the-fisher-lad/)

The works of:

- Ralph Waldo Emmerson
- Paul Theroux

- Robert Frost
- My Side of the Mountain
- By: Jean Craighead George

MARY REYNOLDS' RECOMMENDED BOOKS:

- *Natures Best: A New Approach to Conservation that Starts in Your Yard* by Doug W Tallamy
- *Wilding: Returning Nature to Our Farm* by Isabella Tree
- *Rebirding: Rewilding Britain and its Birds* by Benedict MacDonald
- Buy Mary's book, *The Garden Awakening.*

BURREN NATURE SANCTUARY ECO TOURISM POLICY

We aim to engage our visitors with nature, to foster respect and responsibility and run our business sustainably. We are committed to raising awareness about sustainability issues among staff, suppliers and visitors. We have an ongoing training programme for our staff on sustainability, the natural history of the Burren and our environmental policy.

WE ARE 100% COMMITTED TO:

- Environmental good practice and related principles.
- Monitoring and measuring our impacts.
- Respecting the Geopark Sustainable Code of Practice.
- Cooperating with others to ensure social and environmental benefits for local communities visitors and the region itself.

IN OUR SANCTUARY, WE MAKE USE OF:

- An organic polytunnels.
- Waste composting.
- Eco-friendly construction of our buildings with high insulation.
- An Air-to-water heat pump.
- A Rainwater harvesting system.
- A High spec waste water system with a sand filter.
- Twenty-five (25) acres of land dedicated to conservation and the wildlife sanctuary.

We pride ourselves on being an equal opportunities employer and strive towards inclusiveness and accessibility in all areas.

It is our policy to meet targets in our responsible tourism plan.

IN 2019, WE MET OUR TARGETS TO:

- Increase recyclable waste by 10%.
- Reduce energy by 10%.
- Reduce treated water consumption by 10%.
- Increase green purchasing by 10%.
- Increase the use of sustainable transport by 10%.

WE ARE PROUD TO BE A MEMBER OF THE FOLLOWING ORGANISATIONS:

- Burren Ecotourism.
- Leave no Trace Ireland.
- Burren and Cliffs of Moher Geo Park.
- The All Ireland Pollinator Plan
- Approved Organic Standard IOFGA.

WATER:

- We aim to reduce the consumption of treated water per visitor.
- We will continue our own attempts at water conservation in the kitchen.

WASTE:

- We have a 'by weight' billing system and weigh all kitchen waste that goes to composting, pigs and chickens.
- We monitor portion size and production.
- We use recyclable napkins, food prep gloves and biodegradable J-cloths.
- We have a 'no picnics' policy to reduce waste.

ENERGY:

- Our insulation is very effective. We have replaced all bulbs with LED bulbs (where possible).
- We use more cold water to rinse in the wash up.
- We understand climate change issues and are committed to addressing climate change by meeting our targets.
- We continually plant native trees and hedging to offset carbon usage.

CONTRIBUTION TO CONSERVATION:

- We are proud contributors of 'Adopt a Hedgerow'. Every year, we clear rubbish from Kinvara to our entrance gate (0,62 miles/1 kilometre).

www.ingramcontent.com/pod-product-compliance
Lightning Source LLC
Chambersburg PA
CBHW022110210326
41521CB00028B/180